W0230053

Instrumentation and Process Control

Instrumentation and Process Control

Contributors :
Ruzairi Abdul Rahim,
Mohd Hafiz Fazalul Rahiman, *et al.*

AURIS REFERENCE LTD.
London, UK

Instrumentation and Process Control
Contributors : Ruzairi Abdul Rahim *and* Mohd Hafiz Fazalul Rahiman, *et al.*

Auris Reference Ltd., UK

www.aurisreference.com

United Kingdom

Copyright 2016

Printed in 2017 for Sale in the Indian Subcontinent

The information in this book has been obtained from highly regarded resources. The copyrights for individual articles remain with the authors, as indicated. All chapters are distributed under the terms of the Creative Commons Attribution License, which permit unrestricted use, distribution, and reproduction in any medium, provided the original author and source are credited.

Notice

Contributors, whose names have been given on the book cover, are not associated with the Publisher. The editors and the Publisher have attempted to trace the copyright holders of all material reproduced in this publication and apologise to copyright holders if permission has not been obtained. If any copyright holder has not been acknowledged, please write to us so we may rectify.

Reasonable efforts have been made to publish reliable data. The views articulated in the chapters are those of the individual contributors, and not necessarily those of the editors or the Publisher. Editors and/or the Publisher are not responsible for the accuracy of the information in the published chapters or consequences from their use. The Publisher accepts no responsibility for any damage or grievance to individual(s) or property arising out of the use of any material(s), instruction(s), methods or thoughts in the book.

No part of this publication maybe reproduced, stored in a retrieval system or transmitted in any form or by any means, electronic, mechanical, photocopying, recording, scanning or otherwise without prior written permission of the publisher.

Instrumentation and Process Control

ISBN: 978-1-78154-506-5

British Library Cataloguing in Publication Data
A CIP record for this book is available from the British Library

Exclusively distributed by CBS Publishers & Distributors Pvt. Ltd.

Sales & Distribution Rights only for India, Pakistan, Bangladesh, Sri Lanka, Nepal and Bhutan.This book is not to be sold outside these territories.

PREFACE

Instrumentation is the use of measuring instruments to monitor and control a process. It is the art and science of measurement and control of process variables within a production, laboratory, or manufacturing area. Measurement and control is the brain and nervous system of any modern plant. Measurement and control systems monitor and regulate processes that otherwise would be difficult to operate efficiently and safely while meeting the requirements for high quality and low cost.

Process Instrumentation and Control is needed in modern industrial processes for a business to remain profitable. It improves product quality, reduces plant emissions, minimizes human error, and reduces operating costs among many other benefits.

The role of process control has changed throughout the years and is continuously shaped by technology. The traditional role of process control in industrial operations was to contribute to safety, minimized environmental impact, and optimize processes by maintaining process variable near the desired values. Generally, anything that requires continuous monitoring of an operation involves the role of a process engineer. Instrumentation engineers are commonly responsible for integrating the sensors with the recorders, transmitters, displays or control systems. They may design or specify installation, wiring and signal conditioning. They may be responsible for calibration, testing and maintenance of the system.

This page left intentionally blank.

CONTENTS

Ruzairi Abdul Rahim, Mohd Hafiz Fazalul Rahiman, Leong Lai Chen,
Chan Kok San and Pang Jon Fea

This page left intentionally blank.

LIST OF CONTRIBUTORS

Ruzairi Abdul Rahim

Control & Instrumentation Engineering Department, Faculty of Electrical Engineering, Universiti Teknologi Malaysia, 81310 Skudai, Johor Bahru, Malaysia

Mohd Hafiz Fazalul Rahiman

School of Mechatronic Engineering, Universiti Malaysia Perlis, 02600 Arau, Perlis, Malaysia; E-Mail: hafiz@unimap.edu.my

Leong Lai Chen

Control & Instrumentation Engineering Department, Faculty of Electrical Engineering, Universiti Teknologi Malaysia, 81310 Skudai, Johor Bahru, Malaysia

Chan Kok San

Control & Instrumentation Engineering Department, Faculty of Electrical Engineering, Universiti Teknologi Malaysia, 81310 Skudai, Johor Bahru, Malaysia

Pang Jon Fea

Control & Instrumentation Engineering Department, Faculty of Electrical Engineering, Universiti Teknologi Malaysia, 81310 Skudai, Johor Bahru, Malaysia

This page left intentionally blank.

Chapter 1

INTRODUCTION TO INSTRUMENTATION AND PROCESS CONTROL

INSTRUMENTATION

Instrumentation is the use of measuring instruments to monitor and control a process. It is the art and science of measurement and control of process variables within a production, laboratory, or manufacturing area.

An instrument is a device that measures a physical quantity such as flow, temperature, level, distance, angle, or pressure. Instruments may be as simple as direct reading thermometers or may be complex multi-variable process analyzers. Instruments are often part of a control system in refineries, factories, and vehicles. The control of processes is one of the main branches of applied instrumentation. Instrumentation can also refer to handheld devices that measure some desired variable. Diverse handheld instrumentation is common in laboratories, but can be found in the household as well. For example, a smoke detector is a common instrument found in most western homes.

Instruments attached to a control system may provide signals used to operate solenoids, valves, regulators, circuit breakers, or relays. These devices control a desired output variable, and provide either remote or automated control capabilities. These are often referred to as final control elements when controlled remotely or by a control system.

A transmitter is a device that produces an output signal, often in the form of a 4–20 mA electrical current signal, although many other options using voltage, frequency, pressure, or ethernet are possible. This signal can be used for informational purposes, or it can be sent to a PLC, DCS, SCADA system, LabVIEW or other type of computerized controller, where it can be interpreted into readable values and used to control other devices and processes in the system.

Control instrumentation plays a significant role in both gathering information from the field and changing the field parameters, and as such are a key part of control loops.

Fig. : A control post of a steam turbine.

Elements of Industrial Instrumentation

Elements of industrial instrumentation have long histories. Scales for comparing weights and simple pointers to indicate position are ancient technologies. Some of the earliest measurements were of time. One of the oldest water clocks was found in the tomb of the Egyptian pharaoh Amenhotep I, buried around 1500 BCE. Improvements were incorporated in the clocks. By 270 BCE they had the rudiments of an automatic control system device. In 1663 Christopher Wren presented the Royal Society with a design for a "weather clock". A drawing shows meteorological sensors moving pens over paper driven by clockwork. Such devices did not become standard in meteorology for two centuries. The concept has remained virtually unchanged as evidenced by pneumatic chart recorders, where a pressurized bellows displaces a pen. Integrating sensors, displays, recorders and controls was uncommon until the industrial revolution, limited by both need and practicality.

In the early years of process control, process indicators and control elements such as valves were monitored by an operator that walked around the unit adjusting the valves to obtain the desired temperatures, pressures, and flows. As technology evolved pneumatic controllers were invented and mounted in the field that monitored the process and controlled the valves. This reduced the amount of time process operators were needed to monitor the process. Later years the actual controllers were moved to a central room and signals were sent into the control room to monitor the process and outputs signals were sent to the final control element such as a valve to adjust the process as needed. These controllers and indicators were mounted on a wall called a control board. The operators stood in front of this board walking back and forth monitoring the process indicators. This again reduced the number and amount of time process operators were needed to walk around the units. The most standard pneumatic signal level used during these years was 3-15 psig.

Electronics enabled wiring to replace pipes. The transistor was commercialized by the mid-1950s. Each instrument company introduced their own standard instrumentation signal, causing confusion until the 4-20 mA range was used as the standard electronic instrument signal for transmitters and valves. This signal was eventually standardized as ANSI/ISA S50, "Compatibility of Analog Signals for Electronic Industrial Process Instruments", in the 1970s. The transformation of instrumentation from mechanical pneumatic transmitters, controllers, and valves to electronic instruments reduced maintenance costs as electronic instruments were more dependable than mechanical instruments. This also increased efficiency and production due to their increase in accuracy. Pneumatics enjoyed some advantages, being favored in corrosive and explosive atmospheres.

The pneumatic and electronic signaling standards allowed centralized monitoring and control of a distributed process. The concept was limited by communication line lengths (perhaps 100 meters for pneumatics). Each pipe or wire pair carried one signal. The next evolution of instrumentation came with the production of Distributed Control Systems (DCS) which allowed monitoring and control from multiple locations which could be widely separated. A process operator could sit in front of a screen (no longer a control board) and monitor thousands of points throughout a large complex. A closely related development was termed "Supervisory Control and Data Acquisition" (SCADA). These technologies were supported by personal computers, networks and graphical user interfaces.

Definition

The design, construction, and provision of instruments for measurement, control, etc; the state of being equipped with or controlled by such instruments collectively. It notes that this use of the word originated in the U.S.A. in the early

20th century. More traditional uses of the word were associated with musical or surgical instruments. While the word is traditionally a noun, it is also used as an adjective (as instrumentation engineer, instrumentation amplifier and instrumentation system).

Measurement instruments have three traditional classes of use:

- Monitoring of processes and operations
- Control of processes and operations
- Experimental engineering analysis

While these uses appear distinct, in practice they are less so. All measurements have the potential for decisions and control. A home owner may change a thermostat setting in response to a utility bill computed from meter readings.

Examples

In some cases the sensor is a very minor element of the mechanism. Digital cameras and wristwatches might technically meet the loose definition of instrumentation because they record and/or display sensed information. Under most circumstances neither would be called instrumentation, but when used to measure the elapsed time of a race and to document the winner at the finish line, both would be called instrumentation.

Household

A very simple example of an instrumentation system is a mechanical thermostat, used to control a household furnace and thus to control room temperature. A typical unit senses temperature with a bi-metallic strip. It displays temperature by a needle on the free end of the strip. It activates the furnace by a mercury switch. As the switch is rotated by the strip, the mercury makes physical (and thus electrical) contact between electrodes.

Another example of an instrumentation system is a home security system. Such a system consists of sensors (motion detection, switches to detect door openings), simple algorithms to detect intrusion, local control (arm/disarm) and remote monitoring of the system so that the police can be summoned. Communication is an inherent part of the design.

Kitchen appliances use sensors for control.

- A refrigerator maintains a constant temperature by measuring the internal temperature.
- A microwave oven sometimes cooks *via* a heat-sense-heat-sense cycle until sensing done.
- An automatic ice machine makes ice until a limit switch is thrown.
- Pop-up bread toasters can operate by time or by heat measurements.

- Some ovens use a temperature probe to cook until a target internal food temperature is reached.

- A common toilet refills the water tank until a float closes the valve. The float is acting as a water level sensor.

Automotive

Modern automobiles have complex instrumentation. In addition to displays of engine rotational speed and vehicle linear speed, there are also displays of battery voltage and current, fluid levels, fluid temperatures, distance traveled and feedbacks of various controls (turn signals, parking brake, headlights, transmission position). Cautions may be displayed for special problems (fuel low, check engine, tire pressure low, door ajar, seat belt unfastened). Problems are recorded so they can be reported to diagnostic equipment. Navigation systems can provide voice commands to reach a destination. Automotive instrumentation must be cheap and reliable over long periods in harsh environments. There may be independent airbag systems which contain sensors, logic and actuators. Anti-skid braking systems use sensors to control the brakes, while cruise control affects throttle position. A wide variety of services can be provided *via* communication links as the OnStar system. Autonomous cars (with exotic instrumentation) have been demonstrated.

Aircraft

Early aircraft had a few sensors. "Steam gauges" converted air pressures into needle deflections that could be interpreted as altitude and airspeed. A magnetic compass provided a sense of direction. The displays to the pilot were as critical as the measurements.

A modern aircraft has a far more sophisticated suite of sensors and displays, which are embedded into avionics systems. The aircraft may contain inertial navigation systems, global positioning systems, weather radar, autopilots, and aircraft stabilization systems. Redundant sensors are used for reliability. A subset of the information may be transferred to a crash recorder to aid mishap investigations. Modern pilot displays now include computer displays including head-up displays.

Air traffic control radar is distributed instrumentation system. The ground portion transmits an electromagnetic pulse and receives an echo (at least). Aircraft carry transponders that transmit codes on reception of the pulse. The system displays aircraft map location, an identifier and optionally altitude. The map location is based on sensed antenna direction and sensed time delay. The other information is embedded in the transponder transmission.

Laboratory Instrumentation

Among the possible uses of the term is a collection of laboratory test equipment controlled by a computer through an IEEE-488 bus (also known as GPIB for General Purpose Instrument Bus or HPIB for Hewlitt Packard Instrument

Bus). Laboratory equipment is available to measure many electrical and chemical quantities. Such a collection of equipment might be used to automate the testing of drinking water for pollutants.

Measurement

Instrumentation is used to measure many parameters (physical values). These parameters include:

• Pressure, either differential or static • Flow • Temperature • Levels of liquids, *etc.* • Density	• Viscosity • Other mechanical properties of materials • Properties of ionising radiation • Frequency • Current	• Voltage • Inductance • Capacitance • Resistivity	• Chemical composition • Chemical properties • Properties of light • Vibration • Weight

Control

In addition to measuring field parameters, instrumentation is also responsible for providing the ability to modify some field parameters. That means the instrument is not only for measuring purposes, but also for changing and modification of the process system, these instruments are generally referred to as actuators. In industries, actuators are used to regulate fluid, control flow, moderate temperatures and open/close electric circuits.

INSTRUMENTATION ENGINEERING

Instrumentation engineering is the engineering specialization focused on the principle and operation of measuring instruments that are used in design and configuration of automated systems in electrical, pneumatic domains *etc.* They typically work for industries with automated processes, such as chemical or manufacturing plants, with the goal of improving system productivity, reliability, safety, optimization, and stability. To control the parameters in a process or in a particular system, devices such as microprocessors, microcontrollers or PLCs are used, but their ultimate aim is to control the parameters of a system.

Instrumentation engineering is loosely defined because the required tasks are very domain dependent. An expert in the biomedical instrumentation of laboratory rats has very different concerns than the expert in rocket instrumentation. Common concerns of both are the selection of appropriate sensors based on size, weight, cost, reliability, accuracy, longevity, environmental robustness and frequency response. Some sensors are literally fired in artillery shells. Others sense thermonuclear explosions until destroyed. Invariably sensor data must be recorded, transmitted or displayed. Recording rates and capacities vary enormously. Transmission can be trivial or can be clandestine, encrypted and low-power in the presence of jamming. Displays can be trivially simple or can require consultation with human factors experts. Control system design varies from trivial to a separate specialty.

Instrumentation engineers are commonly responsible for integrating the sensors with the recorders, transmitters, displays or control systems. They may design or specify installation, wiring and signal conditioning. They may be responsible for calibration, testing and maintenance of the system.

In a research environment it is common for subject matter experts to have substantial instrumentation system expertise. An astronomer knows the structure of the universe and a great deal about telescopes - optics, pointing and cameras (or other sensing elements). That often includes the hard-won knowledge of the operational procedures that provide the best results. For example, an astronomer is often knowledgeable of techniques to minimize temperature gradients that cause air turbulence within the telescope.

PROCESS CONTROL

Process controls is a mixture between the statistics and engineering discipline that deals with the mechanism, architectures, and algorithms for controlling a process. Some examples of controlled processes are:

- Controlling the temperature of a water stream by controlling the amount of steam added to the shell of a heat exchanger.
- Operating a jacketed reactor isothermally by controlling the mixture of cold water and steam that flows through the jacket of a jacketed reactor.
- Maintaining a set ratio of reactants to be added to a reactor by controlling their flow rates.
- Controlling the height of fluid in a tank to ensure that it does not overflow.

To truly understand or solve a design problem it is necessary to understand the key concepts and general terminology.

PROCESS CONTROL BACKGROUND

The role of process control has changed throughout the years and is continuously shaped by technology. The traditional role of process control in industrial operations was to contribute to safety, minimized environmental impact, and optimize processes by maintaining process variable near the desired values. Generally, anything that requires continous monitoring of an operation involve the role of a process engineer. In years past the monitoring of these processes was done at the unit and were maintained locally by operator and engineers. Today many chemical plant have gone to full automation which means that engineers and operators are helped by DCS that communicates with the instruments in the field.

Benefits of Process Control

The benefits of controlling or automating process are in a number of distinct area in the operation of a unit or chemical plant. Safety of workers and the community around a plant is probably concern number one or should be for most engineers as they begin to design their processes. Chemical plants have a great potential to do severe damage if something goes wrong and it is inherent the setup of process control to set boundaries on specific unit so that they don't injure or kill workers or individuals in the community.

The Objectives of Control

A control system is required to perform either one or both task:

Maintain the Process at the Operational Conditions and Set Points

Many processes should work at steady state conditions or in a state in which it satisfies all the benefits for a company such as budget, yield, safety, and other quality objectives. In many real-life situations, a process may not always remain static under these conditions and therefore can cause substantial losses to the process. One of the ways a process can wander away from these conditions is by the system becoming unstable, meaning process variables oscillate from its physical boundaries over a limited time span. An example of this would be a water tank in a heating and cooling process without any drainage and is being constantly filled with water. The water level in the tank will continue to rise and eventually overflow. This uncontrolled system can be controlled simply by adding control valves and level sensors in the tank that can tell the engineer or technician the level of water in the tank. Another way a process can stray away from steady state conditions can be due to various changes in the environmental conditions, such as composition of a feed, temperature conditions, or flow rate.

Transition the Process from one Operational Condition to Another

In real-life situations, engineers may change the process operational conditions for a variety of different reasons, such as customer specifications or environment specifications. Although, transitioning a process from one operational condition to another can be detrimental to a process, it also can be beneficial depending on the company and consumer demands.

Examples of why a process may be moved from one operational set point to another:

1. Economics
2. Product specifications
3. Operational constraints
4. Environmental regulations
5. Consumer/Customer specifications
6. Environmental regulations
7. Safety precautions

Definitions and Terminology

In controlling a process there exist two type of classes of variables.

1. **Input Variable –** This variable shows the effect of the surroundings on the process. It normally refers to those factors that influence the process. An example of this would be the flow rate of the steam through a heat exchanger that would change the amount of energy put into the process. There are effects of the surrounding that are controllable and some that are not. These are broken down into two types of inputs.

    a. *Manipulated inputs:* variable in the surroundings can be control by an operator or the control system in place.

    b. *Disturbances:* inputs that can not be controlled by an operator or control system. There exist both measurable and immeasurable disturbances.

2. **Output variable–** Also known as the *control variable* These are the variables that are process outputs that effect the surroundings. An example of this would be the amount of CO_2 gas that comes out of a combustion reaction. These variables may or may not be measured.

    As we consider a controls problem. We are able to look at two major control structures.

1. **Single input-**Single Output (SISO)- for one control(output) varible there exist one manipulate (input) variable that is used to affect the process

2. **Multiple input-**Multiple output(MIMO)- There are several control (output) variable that are affected by several manipulated (input) variables used in a given process.

    - **Cascade:** A control system with 2 or more controllers, a "Master" and "Slave" loop. The output of the "Master" controller is the setpoint for the "Slave" controller.

    - **Dead Time:** The amount of time it takes for a process to start changing after a disturbance in the system.

    - **Derivative Control:** The "D" part of a PID controller. With derivative action the controller output is proportional to the rate of change of the process variable or error.*

    - **Error:** In process controls, error is defined as: Error = setpoint - process variable.

    - **Integral Control:** The "I" part of a PID controller. With integral action the controller output is proportional to the amount and duration of the error signal.

    - **PID Controller:** PID controllers are designed to eliminate the need for continuous operator attention. They are used to automatically adjust system variables to hold a process variable at a setpoint. Error is defined above as the difference between setpoint and process variable.

- **Proportional Control:** The "P" part of a PID controller. With proportional action the controller output is proportional to the amount of the error signal.

- **Setpoint:** The setpoint is where you would like a controlled process variable to be.

Design Methodology for Process Control

1. **Understand the process:** Before attempting to control a process it is necessary to understand how the process works and what it does.

2. **Identify the operating parameters:** Once the process is well understood, operating parameters such as temperatures, pressures, flow rates, and other variables specific to the process must be identified for its control.

3. **Identify the hazardous conditions:** In order to maintain a safe and hazard-free facility, variables that may cause safety concerns must be identified and may require additional control.

4. **Identify the measurables:** It is important to identify the measurables that correspond with the operating parameters in order to control the process.

Measurables for process systems include:

- Temperature
- Pressure
- Flow rate
- pH
- Humidity
- Level
- Concentration
- Viscosity
- Conductivity
- Turbidity
- Redox/potential
- Electrical behavior
- Flammability

5. **Identify the points of measurement:** Once the measurables are identified, it is important locate where they will be measured so that the system can be accurately controlled.

6. **Select measurement methods:** Selecting the proper type of measurement device specific to the process will ensure that the most accurate, stable,

and cost-effective method is chosen. There are several different signal types that can detect different things.

These signal types include:

- Electric
- Pneumatic
- Light
- Radiowaves
- Infrared (IR)
- Nuclear

7. **Select control method:** In order to control the operating parameters, the proper control method is vital to control the process effectively. On/off is one control method and the other is continuous control. Continuous control involves Proportional (P), Integral (I), and Derivative (D) methods or some combination of those three.

8. **Select control system:** Choosing between a local or distributed control system that fits well with the process effects both the cost and efficacy of the overall control.

9. **Set control limits:** Understanding the operating parameters allows the ability to define the limits of the measurable parameters in the control system.

10. **Define control logic:** Choosing between feed-forward, feed-backward, cascade, ratio, or other control logic is a necessary decision based on the specific design and safety parameters of the system.

11. **Create a redundancy system:** Even the best control system will have failure points; therefore it is important to design a redundancy system to avoid catastrophic failures by having back-up controls in place.

12. **Define a fail-safe:** Fail-safes allow a system to return to a safe state after a breakdown of the control. This fail-safe allows the process to avoid hazardous conditions that may otherwise occur.

13. **Set lead/lag criteria:** Depending on the control logic used in the process, there may be lag times associated with the measurement of the operating parameters. Setting lead/lag times compensates for this effect and allow for accurate control.

14. **Investigate effects of changes before/after:** By investigating changes made by implementing the control system, unforeseen problems can be identified and corrected before they create hazardous conditions in the facility.

15. **Integrate and test with other systems:** The proper integration of a new control system with existing process systems avoids conflicts between multiple systems.

STEADY STATE

In systems theory, a system in a steady state has numerous properties that are unchanging in time. This means that for those properties p of the system, the partial derivative with respect to time is zero:

$$\frac{\partial p}{\partial t} = 0$$

The concept of steady state has relevance in many fields, in particular thermodynamics, economics, and engineering. Steady state is a more general situation than dynamic equilibrium. If a system is in steady state, then the recently observed behavior of the system will continue into the future. In stochastic systems, the probabilities that various states will be repeated will remain constant.

In many systems, steady state is not achieved until some time after the system is started or initiated. This initial situation is often identified as a transient state, start-up or warm-up period.

While a dynamic equilibrium occurs when two or more reversible processes occur at the same rate, and such a system can be said to be in steady state, a system that is in steady state may not necessarily be in a state of dynamic equilibrium, because some of the processes involved are not reversible.

For example: The flow of fluid through a tube or electricity through a network could be in a steady state because there is a constant flow of fluid, or electricity. Conversely, a tank being drained or filled with fluid is a system in transient state, because its volume of fluid changes with time.

Applications

Electronics

The electronics, *steady state* is an equilibrium condition of a circuit or network that occurs as the effects of transients are no longer important.

Steady state determination is an important topic, because many design specifications of electronic systems are given in terms of the steady-state characteristics. Periodic steady-state solution is also a prerequisite for small signal dynamic modeling. Steady-state analysis is therefore an indispensable component of the design process.

In some cases, it is useful to consider constant envelope vibration – vibration that never settles down to motionlessness, but continues to move at constant amplitude – a kind of steady-state condition.

Chemistry

In chemistry, thermodynamics, and other chemical engineering, a *steady state* is a situation in which all state variables are constant in spite of ongoing processes that strive to change them. For an entire system to be at steady state, *i.e.*

for all state variables of a system to be constant, there must be a flow through the system (compare mass balance). One of the simplest examples of such a system is the case of a bathtub with the tap open but without the bottom plug: after a certain time the water flows in and out at the same rate, so the water level (the state variable being Volume) stabilizes and the system is at steady state. Of course the Volume stabilizing inside the tub depends on the size of the tub, the diameter of the exit hole and the flowrate of water in. Since the tub can overflow, eventually a steady state can be reached where the water flowing in equals the overflow plus the water out through the drain.

A steady state flow process requires conditions at all points in an apparatus remain constant as time changes. There must be no accumulation of mass or energy over the time period of interest. The same mass flow rate will remain constant in the flow path through each element of the system. Thermodynamic properties may vary from point to point, but will remain unchanged at any given point.

Electrical Engineering

The ability of an electrical machine or power system to regain its original/ previous state is called Steady State Stability.

The stability of a system refers to the ability of a system to return to its steady state when subjected to a disturbance. As mentioned before, power is generated by synchronous generators that operate in synchronism with the rest of the system. A generator is synchronized with a bus when both of them have same frequency, voltage and phase sequence. We can thus define the power system stability as the ability of the power system to return to steady state without losing synchronism. Usually power system stability is categorized into Steady State, Transient and Dynamic Stability

Steady State Stability studies are restricted to small and gradual changes in the system operating conditions. In this we basically concentrate on restricting the bus voltages close to their nominal values. We also ensure that phase angles between two buses are not too large and check for the overloading of the power equipment and transmission lines. These checks are usually done using power flow studies.

Transient Stability involves the study of the power system following a major disturbance. Following a large disturbance in the synchronous alternator the machine power (load) angle changes due to sudden acceleration of the rotor shaft. The objective of the transient stability study is to ascertain whether the load angle returns to a steady value following the clearance of the disturbance.

The ability of a power system to maintain stability under continuous small disturbances is investigated under the name of Dynamic Stability (also known as small-signal stability). These small disturbances occur due random fluctuations in loads and generation levels. In an interconnected power system, these random variations can lead catastrophic failure as this may force the rotor angle to increase steadily.

Mechanical Engineering

When a periodic force is applied to a mechanical system, it will typically reach steady state after going through some transient behavior. This is often observed in vibrating systems, such as a clock pendulum, but can happen with any type of stable or semi-stable dynamic system. The length of the transient state will depend on the initial conditions of the system. Given certain initial conditions a system may be in steady state from the beginning.

Physiology

Homeostasis is the property of a system that regulates its internal environment and tends to maintain a stable, constant condition. Typically used to refer to a living organism, the concept came from that of milieu interieur that was created by Claude Bernard and published in 1865. Multiple dynamic equilibrium adjustment and regulation mechanisms make homeostasis possible.

Fiber Optics

In fiber optics, "steady state" is a synonym for equilibrium mode distribution.

FURTHER CONTROLLER DESIGN

Essentially all tuning methods are based on a model process for which *optimal* controller settings have been determined. Tuning procedures are thus a matter of matching a real or simulated process to the tuning model.

Largely, the more sophisticated the model, the better the controller performance will be.

The simplest model, already discussed, is the *Zeigler-Nichols delay plus lag model*. This can be used in either *open* or *closed* loop form. The original open loop ZN parameters. In practice the ZN open loop settings are unduly conservative. Higher gains (smaller proportional band) should be used with most processes.

Closed loop tuning, Zeigler-Nichols style, with full amplitude continuous oscillation, is not a satisfactory approach for most actual processes! However, it is perfectly satisfactory for use on a simulated process. Other methods with reduced amplidude oscillation or with the process under partial control are more suitable for use on-line and form the basis of a number of *self tuning controllers*.

Some processes cannot be tuned directly by open loop methods because they do not have a stable steady state. Such systems are said to be *open loop unstable*. A simple example of this is level control where in the absence of some sort of feedback, a vessel will either run dry or overfill. A rather different approach is required for such loops.

The describe tuning methods for single loops. For tuning multiloop systems each loop should be tuned separately, with all other loops *disconnected, i.e.* without control. In general, tunings obtained this way will have to be modified to allow for loop interactions.

Open Loop Methods

Described below are a number of open loop tuning methods. First a couple of correlations are shown based on the ZN open loop response. Then a method for tuning systems with no steady state is described.

Controller Parameter Correlations

A number of alternative models to the delay plus lag have been proposed for both open and closed loop tuning, *e.g.* delay plus two lags, three lags, *etc.* Howver, the Zeigler-Nichols model is in fact perfectly satisfactory, but better controller parameters may be derived for it.

For a model system with delay time T_d and first order lag T_1, the fractional dead time T_f is defined as:

$$T_f = T_d / (T_d + T_1)$$

This is used as parameter to determine settings of dimensionless overall loop gain, the product of controller and process gains:

$$\mu' = \text{controller gain * process gain}$$

and dimensionless integral action time:

$$\mu'_1 = \text{controller reset time} / (T_d + T_1)$$

Settings for PI controllers have been proposed by Lopez and Cianconne and Marlin. Estimates of their values, taken from published graphs, are given in the table below.

These settings appear to be much more satisfactory than the simple Z-N settings which are also shown in the table below.

	Ciancone		Lopez		ZN (open)	
T_f	μ'	μ'_1	μ'	μ'_1	μ'	μ'_1
0	1.1	0.23	5.8	0.4	-	-
0.1	1.1	0.23	5.8	0.5	8.1	0.33
0.2	1.8	0.23	3.1	0.6	3.6	0.66
0.3	1.1	0.72	2.1	0.7	2.1	1.0
0.4	1.0	0.72	1.7	0.8	1.35	1.32
0.5	0.8	0.70	0.91	0.9	0.9	1.65
0.6	0.59	0.67	-	-	0.67	1.98
0.7	0.42	0.60	-	-	0.43	2.31
0.8	0.32	0.53	-	-	0.25	2.64

These settings were obtained by minimising the *integral of the absolute value or square of the error* following a disturbance.

Example of Controller Parameter Correlations

Analysis of the open loop response of a first order process gives the following parameters:

- Delay time = 5
- Time Constant = 20
- Gain = 3.5

Hence evaluate the fractional dead time and from this the controller parameters using the three correlations in the table above.

- Fractional Dead Time = 5 / (20 + 5) = 0.2
- Ciacone Parameters:
    - $\mu' = 1.8$
    - $\mu'_1 = 0.23$
    - Controller gain = 1.8 / 3.5 = 0.514
    - Controller reset = 0.23 * (20 + 5) = 5.75
- Lopez Parameters:
    - $\mu' = 3.1$
    - $\mu'_1 = 0.6$
    - Controller gain = 3.1 / 3.5 = 0.886
    - Controller reset = 0.6 * (20 + 5) = 15
- ZN Open Loop:
    - $\mu' = 3.6$
    - $\mu'_1 = 0.66$
    - Controller gain = 3.6 / 3.5 = 1.029
    - Controller reset = 0.66 * (20 + 5) = 16.5

Systems with no Steady State

These are best tuned using a closed loop procedure. If a step change is used in open loop, then the equivalent time delay is easily identified. However, at the end of the time delay period the process output will rise (or fall) continuously until some physical limit is reached.

The average rate of change during this time, normalised by dividing by the size of the input step, may be taken as the reciprocal of an equivalent first order timeconstant and used to estimate parameters using Z-N or either of the above methods.

A difficulty arises is determining the steady state gain of this kind of process, since it has no steady state. In fact the *time constant* as calculated above already includes a sensitivity which is somewhat equivalent to process gain.

Closed Loop Methods

Now we will move on to show a couple of closed loop methods. First the basic Zeigler Nichols method and then a simple closed loop tuning procedure with damped oscillation.

Closed-Loop Zeigler Nichols

The classic *closed-loop Zeigler Nichols* tuning procedure is to advance the gain of proportional only controller until the process is oscillating continuously at a constant amplitude.

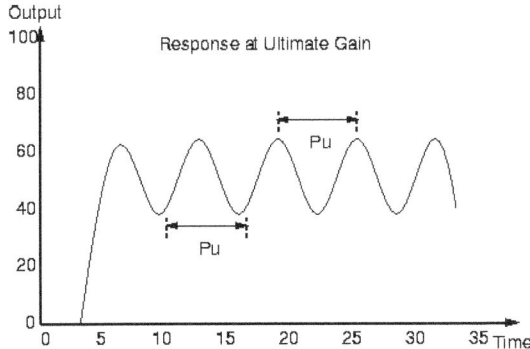

The gain required to achieve this (the *ultimate gain*, μ_u) and the period of oscillation (the *critical period*, P_u) provide parameters from which controller settings are derived as shown below.

Controller Type	Gain	Reset	Derivative
P	$\mu_u/2$	-	-
PI	$\mu_u/2.2$	$P_u/1.2$	-
PID	$\mu_u/1.7$	$P_u/2$	$P_u/8$

Note that it is not necessary here to estimate the steady state process gain as the controller gain, or proportional band, settings are expressed in terms of that actually set on the controller to make the plant oscillate.

Allowing a real process to oscillate between anything but rather tight limits is clearly not likely to be acceptable except for control loops which are noncritical. Two alternatives are therefore used.

- Allow a damped oscillation which dies away as the process approaches a setpoint under control with nonoptimal, but known, controller settings.
- Enforce a limit on the amplitude of oscillation.

The first of these is similar in principle to fitting an open loop model to a delay plus lag, except that it is mathematically a bit more complicated.

The second approach is applied to *self tuning* controllers.

Damped Oscillation Method

Obtain a controller setting which keeps the process within acceptable limits. Use **only proportional action**. Note the gain setting on the controller, μ_d.

Disturb the process with a step change. This is most conveniently done by changing the controller setpoint. The behaviour of the process will be similar to the figure.

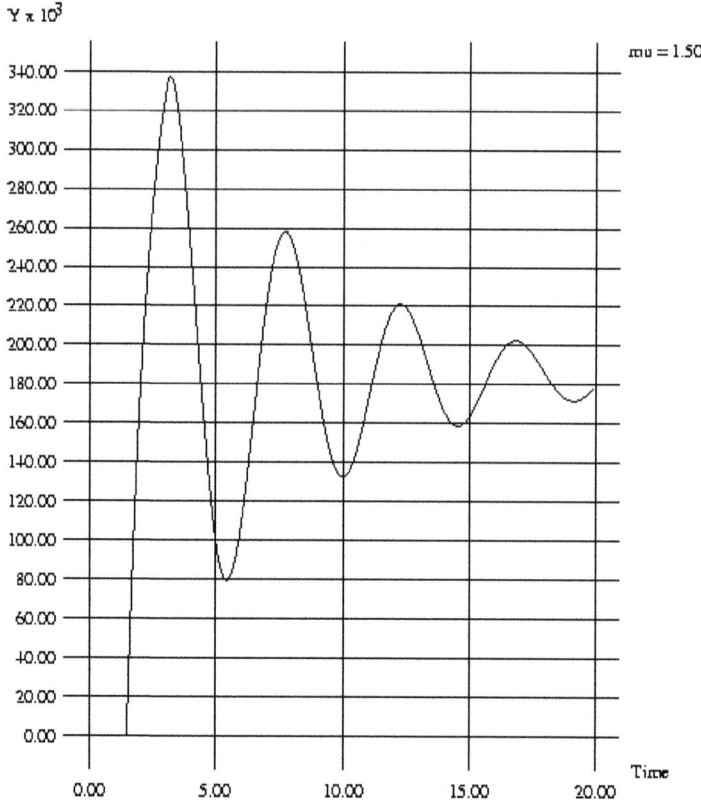

Damped **response of process under non-optimal proportional control**

The process is oscillating rather less than it would if the controller gain were μ_u. It is also oscillating rather more slowly, since increasing the gain of a controller speeds up the response of the process to which it is attached.

The period of oscillation, P_d, e.g. the time between successive peaks or troughs, will thus be slightly greater than P_u, but not much. Hence we can say that typically:

$$P_u \text{ is approximately equal to } 0.95\ P_d$$

We can estimate the increase in gain required to cause the system to oscillate continuously if we know the amount which the oscillation decays between *half cycles*. This would be the ratio between the amplitude of *e.g.* the first peak and that of the trough which follows it. This is not so easy to estimate, as the zero point of the sinusoidal response may not be obvious. Instead take the ratios of peak-to-trough and following trough-to-peak distances. The gain would need to be increased in proportion to this ratio to cause steady oscillation.

Example of Damped Oscillation Method

In the response shown above the proportional controller had a gain of 1.5.

There are 4 complete sinusoids in about 18 time units, so

P_d *is approximately equal to* **18/4 = 4.5**

P_u = 0.95 * 4.5 = 4.28

The first peak to trough distance is about (340-80) = 260 units

The following trough to peak is (80-260) = - 180 units

The ratio of their magnitudes is thus 180/260 = 0.69

The gain should be increased by this factor, i.e:

$$\mu_u = \mu_d/0.69 = 1.5/0.69 = 2.17$$

Hence the ZN PI controller settings are:

Gain = μ_u/2.2 = 2.17/2.2 = 0.99

Reset = P_u/1.2 = 4.28/1.2 = 3.57

Relay Based Tuning Controller

Another way to make a process oscillate is to use a control system which is intrinsically oscillatory. An *on-off* control system has this property, which is why it is seldom used in chemical process control.

However, restricting the size of change which the controller can produce will restrict the amplitude of the oscillations. These can thus be small enough not to upset the process, or its operators!

This controller has a parameter *deadzone*, which defines the minimum change in the measured variable required to produce a change in output. The other parameter defines the size of the output of the controller. This is set to be + *output/2* when the controller is on and - *output/2* when it is off.

The algorithm for the on-off controller is as follows.

error = setpoint - measurement

if (error>deadzone) adjustment = output / 2

if (error<deadzone) adjustment = - output /2

The size of *output* can be quite small, although it must be enough to swing the measured variable across the range of the deadzone. This latter must be larger than any noise in the process.

This controller will oscillate at a period equal to P_u. (Actually if the output is too close to the value required to swing the measurement across the whole deadzone, the period will be slightly longer. It should be enough to cover 4 or 5 times the deadzone.)

If the observed amplitude of the oscillations (maximum to minimum) is a the ultimate gain of the process can be shown to be:

$$\mu_u = (4 * output) / (\pi * a)$$

Example of Relay Based Controller Tuning

We can use the above algorithm in an actual experiment. In the example below the first order process parameters are:

- Delay time = 5
- Time constant = 20
- Gain = 3.5

Other information required by the program is a value for the controller deadzone = 0.01.

Now let us run the experiment with the output of the controller restricted to 4. This gives the following response.

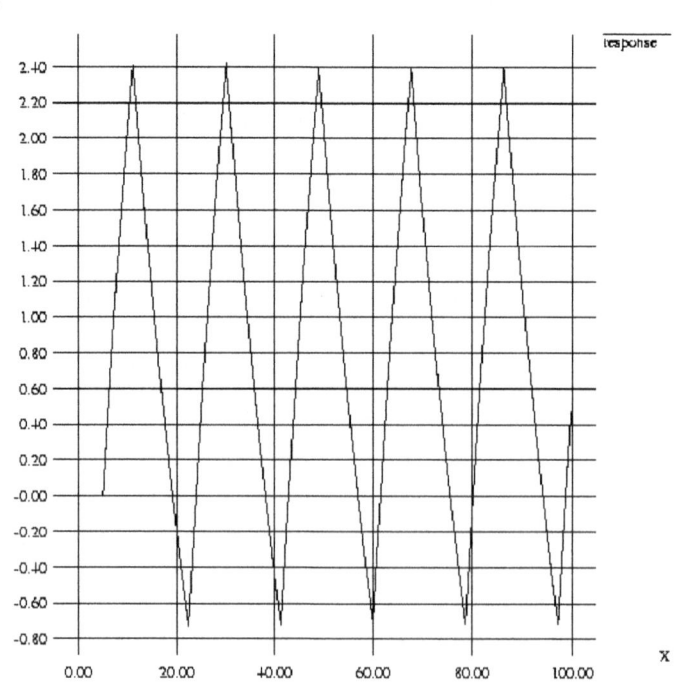

From this response we can determine the value of a, the observed amplitude of the oscillations and the period of oscillation.

- a = 3.1
- Pu = 19

  and hence evaluate the ultimate gain and the tuning parameters.

- $\mu_u = (4 * 4)/(\pi * 3.1) = 1.64$
- Controller gain = 0.747
- Controller reset = 15.83

Now let's try the same process again this time restricting the controller output to 10.

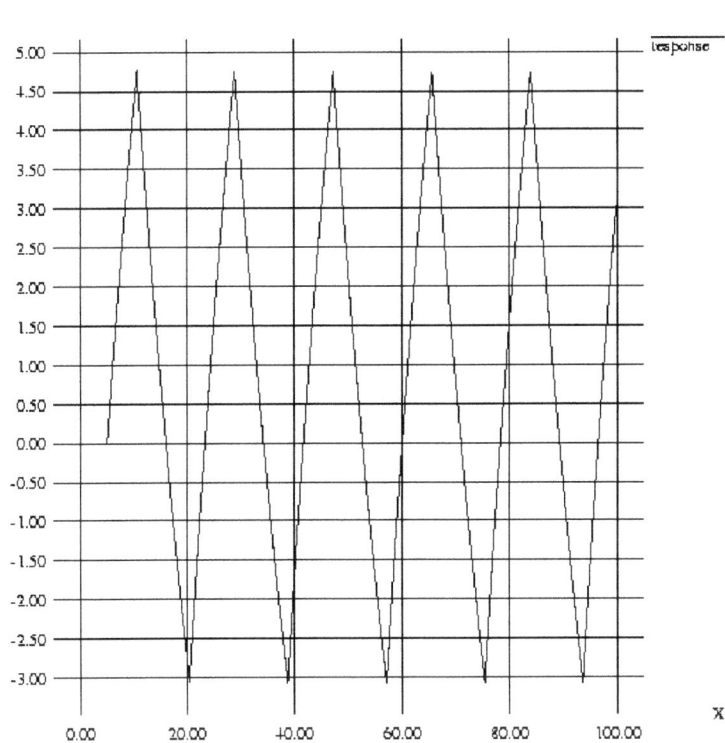

From this response we can determine the value of a, the observed amplitude of the oscillations and the period of oscillation.

- a = 7.75
- Pu = 19

  and hence evaluate the ultimate gain and the tuning parameters.

- $\mu_u = (4 * 10)/(\pi * 7.75) = 1.64$
- Controller gain = 0.747
- Controller reset = 15.83

As you can see the two examples give the same values for the tuning parameters. How well do these values actually control the process at a setpoint of 0.5?

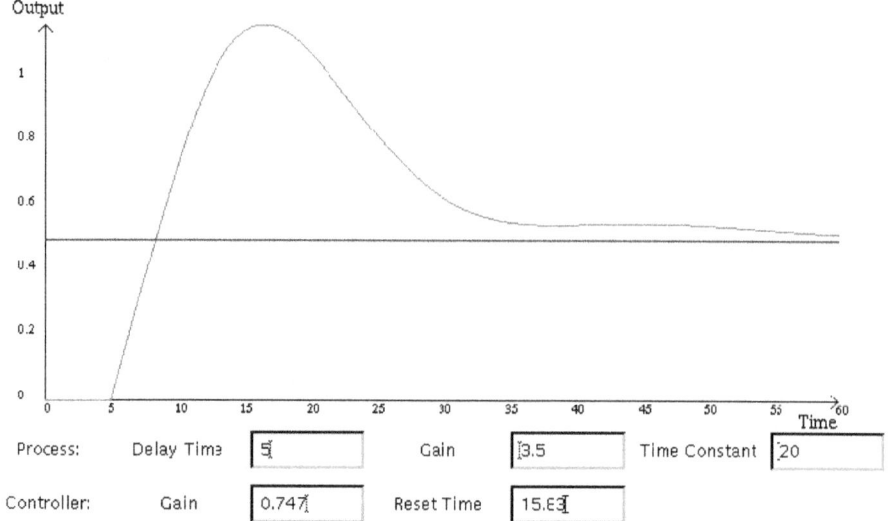

| Process: | Delay Time | 5 | Gain | 3.5 | Time Constant | 20 |
| Controller: | Gain | 0.747 | Reset Time | 15.E3 | | |

This approach can form the basis of a self tuning controller which is switched into on-off or relay mode to determine P_u and μ_u and hence tuning parameters which then get set into the controller.

Real Controllers

Real controllers differ from the ideal ones used in most exercises in several ways:

- Their inputs and outputs are dimensionless (0-100%, 4-20mA, +/- 5V...) although measurements are normally displayed in the control room in appropriate units.
- They work with proportional band rather than gain.
- Derivative action is not implemented as a true derivative.
- Limits are put on the amount of integral action applied by the controller.

Compensated Derivative Action

True derivative action is neither possible nor desirable. If a true derivative controller saw a step change in a measurement it would have to produce an infinite output. Thus a small but sharp noise *spike*, very common on practical electrical measurements, could produce a very large, sudden and spurious, change in a process adjustment. this is obviously highly undesirable, and so the derivative term on any controller must be modified to prevent this happening.

Derivative action is sometimes omitted altogether. In previous tuning examples we have use only PI control. In practice derivative action should be used only when very precise control is required on *measurements which are known to be reliable*. If in doubt, use PI only.

All real controllers use what is called *compensated rate action*. This introduces another parameter which is used to *damp* the derivative action to minimise the effect of noise spikes.

Most controllers nowadays are implemented digitally. To obtain derivatives a computer based controller would require numerical differentiation. This is an unsatisfactory procedure, since it also introduces noise. The rate compensated controller is described entirely by o.d.es, and so is implemented using only numerical *integration*.

Here is the theory of the compensated controller.

We require:

$$\frac{de}{dt} = f(e)$$

Without performing numerical differentiation.

Let:

$$\frac{dz}{dt} = e - \frac{1}{\alpha}z$$

Solve for **z** by numerical *integration*.

Then for small α:

$$\dot{e} \approx \frac{e - \frac{1}{\alpha}z}{\alpha}$$

α is called the rate compensation parameter, and can either have a fixed value or be made, *e.g.* **0.1 * deadtime of process**.

This corresponds to a form of *filtering* of the derivative term which prevents it goint to infinity in the presence of a step disturbance.

Anti-Windup Controller

Consider what will happen if for any reason an error persists in a controller measurement. A simple integral action integrator would just keep on building up to a larger and larger value. If later the measurement error is removed, the *integral* is still there and will take about as long as the original error persisted before it *unwinds* back to a sensible value.

To prevent this happening the size of the integral has to be limited. Either an arbitrary maximum may be set, or else integration may be *turned off* in certain circumstances. These would typically be when either the measurement or controller output reach the limit of their ranges.

BASIC CONTROLLER

The Basic Controller for an application can be visualized as

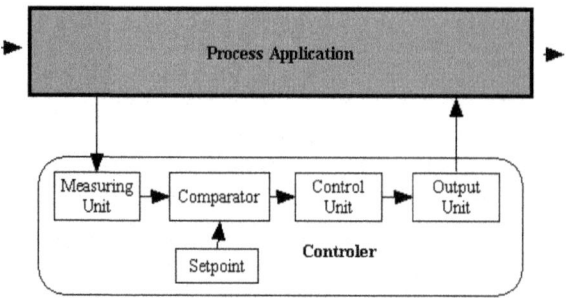

www.engineeringtoolbox.com

The controller consists of :

- a measuring unit with an appropriate instrument to measure the state of process, a temperature transmitter, pressure transmitter or similar
- a input set point device to set the desired value
- a comparator for comparing the measured value with the set point, calculating the difference or error between the two
- a control unit to calculate the output magnitude and direction to compensate the deviation from the desired value
- a output unit converting the output from the controller to physical action, a control valve, a motor or similar

Controller Principles

The Control Units are in general build on the control principles

- proportional controller
- integral controller
- derivative controller

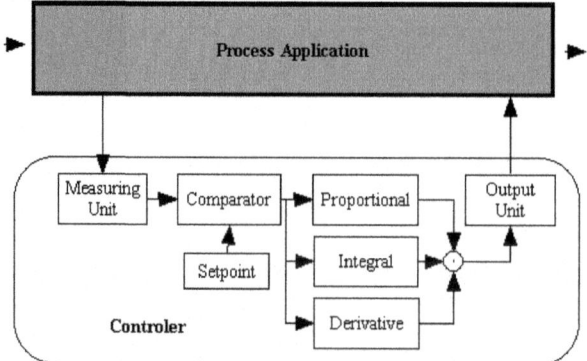

www.engineeringtoolbox.com

Proportional Controller (P-Controller)

One of the most used controllers is the Proportional Controller (P-Controller) who produce an output action that is proportional to the deviation between the set point and the measured process value.

$$O_p = -k_p \, Er$$

where

O_p = *output proportional controller*

k_p = *proportional gain or action factor of the controller*

Er = *error or deviation between the set point value and the measured value*

The gain or action factor - k_p

- influence on the output with a magnitude of kP
- determines how fast the system responds. If the value is too large the system will be in danger to oscillate and/or become unstable. If the value is too small the system error or deviation from set point will be very large
- can be regarded linear only for very small variations

The gain k_p can be expressed as

$$k_p = 100 / P$$

where

P = *proportional band*

The proportional band P, express the value necessary for *100%* controller output. If $P = 0$, the gain or action factor k_p would be infinity - the control action would be *ON/OFF*.

Integral Controller (I-Controller)

With integral action, the controller output is proportional to the amount of time the error is present. Integral action eliminates offset.

$$O_I = - k_I \, \Sigma(Er \, dt)$$

where

O_I = *output integrating controller*

k_I = *integrating gain or action factor of the controller*

dt = *time sample*

The integral controller produce an output proportional to the summarized deviation between the set point and measured value and integrating gain or action factor.

Integral controllers tend to respond slowly at first, but over a long period of time they tend to eliminate errors.

The integral controller eliminates the steady-state error, but may make the transient response worse. The controller may be unstable.

The integral regulator may also cause problems during shutdowns and start up as a result of the integral saturation or wind up effect. An integrating regulator with over time deviation (typical during plant shut downs) will summarize the output to +/- 100%. During start up the output is set to 100%m which may be catastrophic.

Derivative Controller (D-Controller)

With derivative action, the controller output is proportional to the rate of change of the measurement or error. The controller output is calculated by the rate of change of the deviation or error with time.

$O_D = -k_D \, dEr/dt$

where

O_D = *output derivative controller*

k_D = *derivative gain or action factor of the controller*

dEr = *deviation change over time sample dt*

dt = *time sample*

The derivative or differential controller is never used alone. With sudden changes in the system the derivative controller will compensate the output fast. The long term effects the controller allow huge steady state errors.

A derivative controller will in general have the effect of increasing the stability of the system, reducing the overshoot, and improving the transient response.

Proportional, Integral, Derivative Controller (PID-Controller)

The functions of the individual proportional, integral and derivative controllers complements each other. If they are combined its possible to make a system that responds quickly to changes (derivative), tracks required positions (proportional), and reduces steady state errors (integral).

Note that these correlations may not be exactly accurate, because P, I and D are dependent of each other. Changing one of these variables can change the effect of the other two.

Controller Response	Rise Time	Overshoot	Settling Time	Steady State Error
P	Decrease	Increase	Small Change	Decrease
I	Decrease	Increase	Increase	Eliminate
D	Small Change	Decrease	Decrease	Small Change

SYSTEM METRICS

When a system is being designed and analyzed, it doesn't make any sense to test the system with all manner of strange input functions, or to measure all sorts of arbitrary performance metrics. Instead, it is in everybody's best interest to test the system with a set of standard, simple reference functions. Once the system is tested with the reference functions, there are a number of different metrics that we can use to determine the system performance.

It is worth noting that the metrics presented in this chapter represent only a small number of possible metrics that can be used to evaluate a given system.

Standard Inputs

There are a number of standard inputs that are considered simple enough and universal enough that they are considered when designing a system. These inputs are known as a unit step, a ramp, and a parabolic input.

Unit Step

A unit step function is defined piecewise as such:

[Unit Step Function]

$$u(t) = \begin{cases} 0, & t < 0 \\ 1, & t \geq 0 \end{cases}$$

The unit step function is a highly important function, not only in control systems engineering, but also in signal processing, systems analysis, and all branches of engineering. If the unit step function is input to a system, the output of the system is known as the step response.

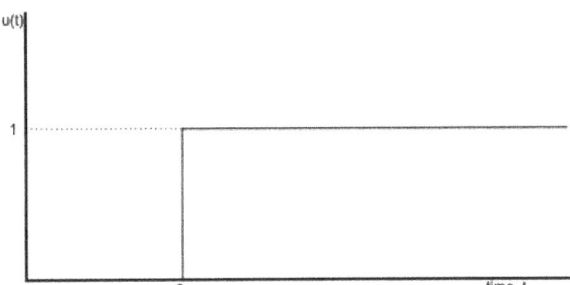

Ramp

A unit ramp is defined in terms of the unit step function, as such:

[Unit Ramp Function]

$$r(t) = tu\,(t)$$

It is important to note that the unit step function is simply the differential of the unit ramp function:

$$r(t) = \int u(t)dt = tu(t)$$

This definition will come in handy when we learn about the Laplace Transform.

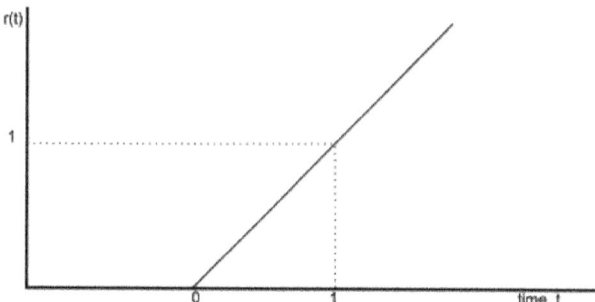

Parabolic

A unit parabolic input is similar to a ramp input:

[Unit Parabolic Function]

$$p(t) = \frac{1}{2}t^2u(t)$$

Notice also that the unit parabolic input is equal to the integral of the ramp function:

$$p(t) = \int r(t)dt = \int tu(t)dt = \frac{1}{2}t^2u(t) = \frac{1}{2}tr(t)$$

Again, this result will become important when we learn about the Laplace Transform.

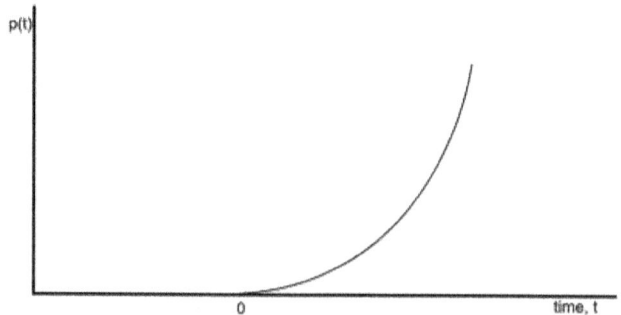

Also, sinusoidal and exponential functions are considered basic, but they are too difficult to use in initial analysis of a system.

Steady State

When a unit-step function is input to a system, the steady-state value of that system is the output value at time $t = \infty$. Since it is impractical (if not completely impossible) to wait till infinity to observe the system, approximations and mathematical calculations are used to determine the steady-state value of the system. Most system responses are asymptotic, that is that the response approaches a particular value. Systems that are asymptotic are typically obvious from viewing the graph of that response.

Step Response

The step response of a system is most frequently used to analyze systems, and there is a large amount of terminology involved with step responses. When exposed to the step input, the system will initially have an undesirable output period known as the transient response. The transient response occurs because a system is approaching its final output value. The steady-state response of the system is the response after the transient response has ended.

The amount of time it takes for the system output to reach the desired value (before the transient response has ended, typically) is known as the rise time. The amount of time it takes for the transient response to end and the steady-state response to begin is known as the settling time.

It is common for a systems engineer to try and improve the step response of a system. In general, it is desired for the transient response to be reduced, the rise and settling times to be shorter, and the steady-state to approach a particular desired "reference" output.

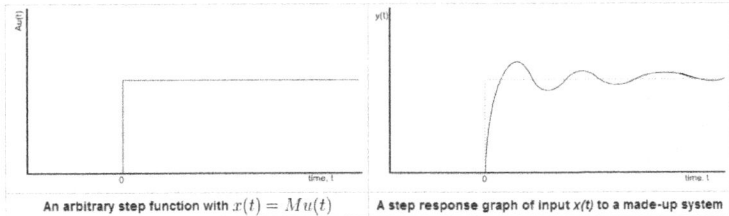

An arbitrary step function with $x(t) = Mu(t)$ A step response graph of input x(t) to a made-up system

Target Value

The target output value is the value that our system attempts to obtain for a given input. This is not the same as the steady-state value, which is the actual value that the target does obtain. The target value is frequently referred to as the reference value, or the "reference function" of the system. In essence, this is the value that we want the system to produce. When we input a "5" into an elevator, we want the output (the final position of the elevator) to be the fifth floor. Pressing the "5" button is the reference input, and is the expected value that we want to obtain. If we press the "5" button, and the elevator goes to the third floor, then our elevator is poorly designed.

Rise Time

Rise time is the amount of time that it takes for the system response to reach the target value from an initial state of zero. Many texts on the subject define the rise time as being the time it takes to rise between the initial position and 80% of the target value. This is because some systems never rise to 100% of the expected, target value, and therefore they would have an infinite rise-time. This book will specify which convention to use for each individual problem. Rise time is typically denoted t_r, or t_{rise}.

Percent Overshoot

Under damped systems frequently overshoot their target value initially. This initial surge is known as the "overshoot value". The ratio of the amount of overshoot to the target steady-state value of the system is known as the percent overshoot. Percent overshoot represents an overcompensation of the system, and can output dangerously large output signals that can damage a system. Percent overshoot is typically denoted with the term PO.

Consider an ordinary household refrigerator. The refrigerator has cycles where it is on and when it is off. When the refrigerator is on, the coolant pump is running, and the temperature inside the refrigerator decreases. The temperature decreases to a much lower level than is required, and then the pump turns off.

When the pump is off, the temperature slowly increases again as heat is absorbed into the refrigerator. When the temperature gets high enough, the pump turns back on. Because the pump cools down the refrigerator more than it needs to initially, we can say that it "overshoots" the target value by a certain specified amount.

Another example concerning a refrigerator concerns the electrical demand of the heat pump when it first turns on. The pump is an inductive mechanical motor, and when the motor first activates, a special counter-acting force known as "back EMF" resists the motion of the motor, and causes the pump to draw more electricity until the motor reaches its final speed. During the startup time for the pump, lights on the same electrical circuit as the refrigerator may dim slightly, as electricity is drawn away from the lamps, and into the pump. This initial draw of electricity is a good example of overshoot.

Steady-State Error

Sometimes a system might never achieve the desired steady-state value, but instead will settle on an output value that is not desired. The difference between the steady-state output value to the reference input value at steady state is called the steady-state error of the system. We will use the variable e_{ss} to denote the steady-state error of the system.

Settling Time

After the initial rise time of the system, some systems will oscillate and vibrate for an amount of time before the system output settles on the final value. The amount of time it takes to reach steady state after the initial rise time is known as the settling time. Notice that damped oscillating systems may never settle completely, so we will define settling time as being the amount of time for the system to reach, and stay in, a certain acceptable range. The acceptable range for settling time is typically determined on a per-problem basis, although common values are 20%, 10%, or 5% of the target value. The settling time will be denoted as t_s.

System Order

The order of the system is defined by the number of independent energy storage elements in the system, and intuitively by the highest degree of the linear differential equation that describes the system. In a transfer function representation, the order is the highest exponent in the transfer function. In a proper system, the system order is defined as the degree of the denominator polynomial. In a state-space equation, the system order is the number of state-variables used in the system. The order of a system will frequently be denoted with an n or N, although these variables are also used for other purposes.

Proper Systems

A proper system is a system where the degree of the denominator is larger than or equal to the degree of the numerator polynomial. A strictly proper system is a system where the degree of the denominator polynomial is larger than (but never equal to) the degree of the numerator polynomial. A biproper system is a system where the degree of the denominator polynomial equals the degree of the numerator polynomial.

It is important to note that only proper systems can be physically realized. In other words, a system that is not proper cannot be built. It makes no sense to spend a lot of time designing and analyzing imaginary systems.

Example: System Order

Find the order of this system:

$$G(s) = \frac{1+s}{1+s+s^2}$$

The highest exponent in the denominator is s^2, so the system is order 2. Also, since the denominator is a higher degree than the numerator, this system is strictly proper.

In the above example, G(s) is a second-order transfer function because in the denominator one of the s variables has an exponent of 2. Second-order functions are the easiest to work with.

System Type

Let's say that we have a process transfer function (or combination of functions, such as a controller feeding in to a process), all in the forward branch of a unity feedback loop. Say that the overall forward branch transfer function is in the following generalized form (known as pole-zero form):

[Pole-Zero Form]

$$G(s) = \frac{K \prod_i (s - s_i)}{s^M \prod_j (s - s_j)}$$

Poles at the origin are called integrators, because they have the effect of performing integration on the input signal.

we call the parameter M the system type. Note that increased system type number correspond to larger numbers of poles at $s = 0$. More poles at the origin generally have a beneficial effect on the system, but they increase the order of the system, and make it increasingly difficult to implement physically. System type will generally be denoted with a letter like N, M, or m. Because these variables are typically reused for other purposes, this book will make clear distinction when they are employed.

Now, we will define a few terms that are commonly used when discussing system type. These new terms are Position Error, Velocity Error, and Acceleration Error. These names are throwbacks to physics terms where acceleration is the derivative of velocity, and velocity is the derivative of position. Note that none of these terms are meant to deal with movement, however.

Position Error

The position error, denoted by the position error constant K_p. This is the amount of steady-state error of the system when stimulated by a unit step input. We define the position error constant as follows:

[Position Error Constant]

$$K_p = \lim_{s \to 0} G(s)$$

Where G(s) is the transfer function of our system.

Velocity Error

The velocity error is the amount of steady-state error when the system is stimulated with a ramp input. We define the velocity error constant as such:

[Velocity Error Constant]

$$K_v = \lim_{s \to 0} sG(s)$$

Acceleration Error

The acceleration error is the amount of steady-state error when the system is stimulated with a parabolic input. We define the acceleration error constant to be:

[Acceleration Error Constant]

$$K_a = \lim_{s \to 0} s^2 G(s)$$

Now, this table will show briefly the relationship between the system type, the kind of input (step, ramp, parabolic), and the steady-state error of the system:

	Unit System Input		
Type, M	**Au(t)**	**Ar(t)**	**Ap(t)**
0	$e_{ss} = \dfrac{A}{1 + K_p}$	$e_{ss} = \infty$	$e_{ss} = \infty$
1	$e_{ss} = 0$	$e_{ss} = \dfrac{A}{K_v}$	$e_{ss} = \infty$
2	$e_{ss} = 0$	$e_{ss} = 0$	$e_{ss} = \dfrac{A}{K_a}$
> 2	$e_{ss} = 0$	$e_{ss} = 0$	$e_{ss} = 0$

Z-Domain Type

Likewise, we can show that the system order can be found from the following generalized transfer function in the Z domain:

$$G(z) = \frac{K \prod_i (z - z_i)}{(z-1)^M \prod_j (z - z_j)}$$

Where the constant M is the order of the digital system. Now, we will show how to find the various error constants in the Z-Domain:

[Z-Domain Error Constants]

Error Constant	**Equation**
Kp	$K_p = \lim_{z \to 1} G(z)$
Kv	$K_v = \lim_{z \to 1} (z - 1) G(z)$
Ka	$K_a = \lim_{z \to 1} (z - 1)^2 G(z)$

Visually

The various system metrics, acting on a system in response to a step input:

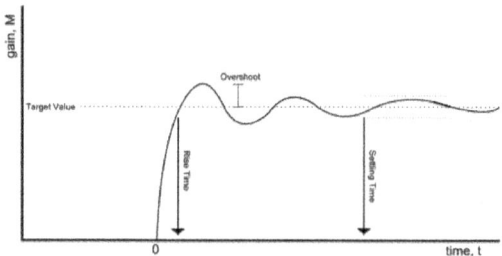

The target value is the value of the input step response. The rise time is the time at which the waveform first reaches the target value. The overshoot is the amount by which the waveform exceeds the target value. The settling time is the time it takes for the system to settle into a particular bounded region. This bounded region is denoted with two short dotted lines above and below the target value.

TRANSIENT STATE AND STEADY STATE RESPONSE OF CONTROL SYSTEM

Under Control System

When we study the analysis of the **transient state and steady state response of control system** it is very essential to know a few basic terms and these are described below.

Standard Input Signals : These are also known as test input signals. The input signal is very complex in nature, it is complex because it may be a combination of various other signals. Thus it is very difficult to analyze characteristic performance of any system by applying these signals. So we use test signals or standard input signals which are very easy to deal with. We can easily analyze the characteristic performance of any system more easily as compared to non standard input signals. Now there are various types of standard input signals and they are written below:

Unit Impulse Signal : In the time domain it is represented by $\partial(t)$. The Laplace transformation of unit impulse function is 1 and the corresponding waveform associated with the unit impulse function is shown below.

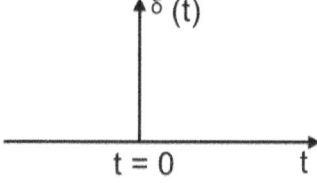

Unit Step Signal : In the time domain it is represented by u (t). The Laplace transformation of unit step function is 1/s and the corresponding waveform associated with the unit step function is shown below.

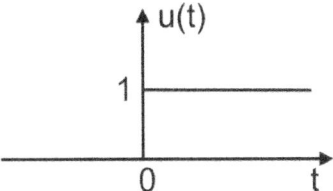

Unit Ramp signal : In the time domain it is represented by r (t). The Laplace transformation of unit ramp function is $1/s^2$ and the corresponding waveform associated with the unit ramp function is shown below.

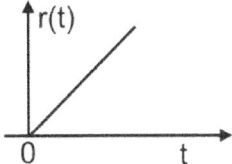

Parabolic Type Signal : In the time domain it is represented by $t^2 / 2$. The Laplace transformation of parabolic type of the function is $1 / s^3$ and the corresponding waveform associated with the parabolic type of the function is shown below.

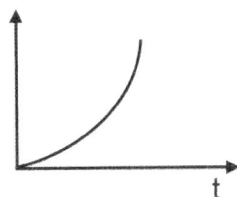

Sinusoidal Type Signal : In the time domain it is represented by sin (ωt). The Laplace transformation of sinusoidal type of the function is $\omega / (s^2 + \omega^2)$ and the corresponding waveform associated with the sinusoidal type of the function is shown below.

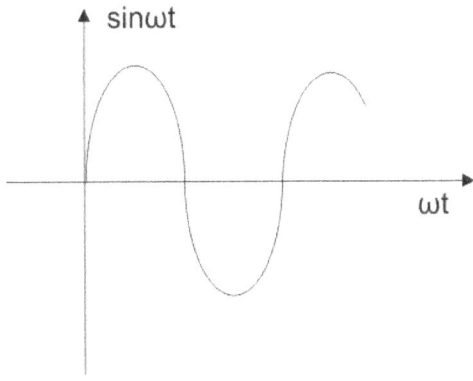

Cosine Type of Signal : In the time domain it is represented by cos (ωt). The Laplace transformation of the cosine type of the function is $\omega \,/\, (s^2 + \omega^2)$ and the corresponding waveform associated with the cosine type of the function is shown below.

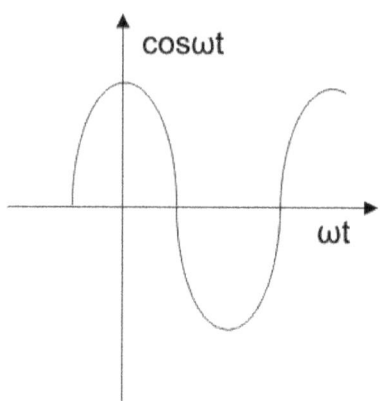

Now are in a position to describe the two types of responses which are a function of time.

Transient Response of Control System

As the name suggests **transient response of control system** means changing so, this occurs mainly after two conditions and these two conditions are written as follows-

- **Condition one :** Just after switching 'on' the system that means at the time of application of an input signal to the system.
- **Condition second :** Just after any abnormal conditions. Abnormal conditions may include sudden change in the load, short circuiting *etc.*

Steady State Response of Control System

Steady state occurs after the system becomes settled and at the steady system starts working normally. **Steady state response of control system** is a function of input signal and it is also called as forced response.

Now the transient state response of control system gives a clear description of how the system functions during **transient state and steady state response of control system** gives a clear description of how the system functions during steady state. Therefore the time analysis of both states is very essential. We will separately analyze both the types of responses. Let us first analyze the transient response. In order to analyze the transient response, we have some time specifications and they are written as follows:

Delay Time: This time is represented by t_d. The time required by the response to reach fifty percent of the final value for the first time, this time is known as delay time. Delay time is clearly shown in the time response specification curve.

Rise Time : This time is represented by t_r. We define rise time in two cases:

1. In case of under damped systems where the value of ζ is less than one, in this case rise time is defined as the time required by the response to reach from zero value to hundred percent value of final value.

2. In case of over damped systems where the value of ζ is greater than one, in this case rise time is defined as the time required by the response to reach from ten percent value to ninety percent value of final value.

Peak Time: This time is represented by t_p. The time required by the response to reach the peak value for the first time, this time is known as peak time. Peak time is clearly shown in the time response specification curve.

Settling Time : This time is represented by t_s. The time required by the response to reach and within the specified range of about (two percent to five percent) of its final value for the first time, this time is known as settling time. Settling time is clearly shown in the time response specification curve.

Maximum Overshoot: It is expressed (in general) in percentage of the steady state value and it is defined as the maximum positive deviation of the response from its desired value. Here desired value is steady state value.

Steady State Error : It can be defined as the difference between the actual output and the desired output as time tends to infinity.

Now we are in position we to do a time response analysis of a first order system.

Transient State and Steady State Response of First Order Control System

Let us consider the block diagram of the first order system.

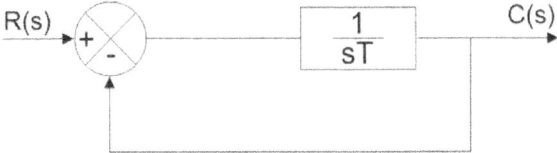

From this block diagram we can find overall transfer function which is linear in nature. The transfer function of the first order system is $1/((sT+1))$. We are going to analyze the steady state and transient response of control system for the following standard signal.

1. Unit impulse.
2. Unit step.
3. Unit ramp.

Unit impulse response : We have Laplace transform of the unit impulse is 1. Now let us give this standard input to a first order system, we have

$$C(s) = \frac{1}{1+sT}$$

Now taking the inverse Laplace transform of the above equation, we have

$$C(t) = \frac{e^{-t/T}}{T}$$

It is clear that the **steady state response of control system** depends only on the time constant 'T' and it is decaying in nature.

Unit step response : We have Laplace transform of the unit impulse is $1/s$. Now let us give this standard input to first order system, we have

$$C(s) = \frac{1}{s(1+sT)}$$

With the help of partial fraction, taking the inverse Laplace transform of the above equation, we have

$$c(t) = 1 - e^{-t/T}$$

It is clear that the time response depends only on the time constant 'T'. In this case the steady state error is zero by putting the limit t is tending to zero.

Unit ramp response : We have Laplace transform of the unit impulse is $1/s^2$. Now let us give this standard input to first order system, we have

$$C(s) = \frac{1}{s^2(1+sT)}$$

With the help of partial fraction, taking the inverse Laplace transform of the above equation we have

$$c(t) = 1 - T + Te^{-t/T}$$

On plotting the exponential function of time we have 'T' by putting the limit t is tending to zero.

Transient State and Steady State Response of Second Order Control System

Let us consider the block diagram of the second order system.

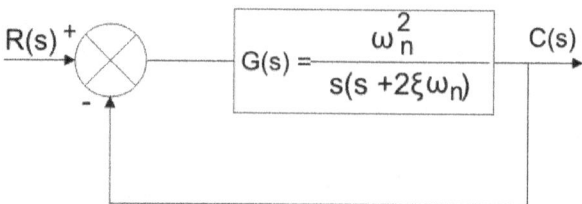

From this block diagram we can find overall transfer function which is non-linear in nature. The transfer function of the second order system is

$$(\omega^2) / (s(s + 2\zeta\omega)).$$

We are going to analyze the **transient state response of control system** for the following standard signal.

Unit impulse response : We have Laplace transform of the unit impulse is 1. Now let us give this standard input to second order system, we have

$$C(S) = \frac{\omega^2}{s(s + 2\omega\zeta)}$$

Where ω is natural frequency in rad/sec and ζ is damping ratio.

Unit step response : We have Laplace transform of the unit impulse is $1/s$. Now let us give this standard input to first order system, we have

$$C(S) = \frac{\omega^2}{s(s + 2\omega\zeta)}$$

With the help of partial fraction, taking the inverse Laplace transform of the above equation we have

$$c(t) = 1 - \frac{e^{-\zeta\omega t} \sin\left[\omega\sqrt{1 - \zeta^2}\, t + \tan^{-1} \frac{\sqrt{1 - \zeta^2}}{\zeta}\right]}{\sqrt{1 - \zeta^2}}$$

Now we will see the effect of different values of ζ on the response. We have three types of systems on the basis of different values of ζ.

1. **Under damped system :** A system is said to be under damped system when the value of ζ is less than one. In this case roots are complex in nature and the real parts are always negative. System is asymptotically stable. Rise time is lesser than the other system with the presence of finite overshoot.

2. **Critically damped system :** A system is said to be critically damped system when the value of ζ is one. In this case roots are real in nature and the real parts are always repetitive in nature. System is asymptotically stable. Rise time is less in this system and there is no presence of finite overshoot.

3. **Over damped system :** A system is said to be over damped system when the value of ζ is greater than one. In this case roots are real and distinct in nature and the real parts are always negative. System is asymptotically stable. Rise time is greater than the other system and there is no presence of finite overshoot.

4. **Sustained Oscillations :** A system is said to be sustain damped system when the value of zeta is zero. No damping occurs in this case.

Now let us derive the expressions for rise time, peak time, maximum overshoot, settling time and steady state error with a unit step input for second order system.

Rise time : In order to derive the expression for the rise time we have to equate the expression for c(t) = 1. From the above we have

$$c(t) = 1 = 1 - \frac{e^{-\zeta \omega t} \sin\left[\omega\sqrt{1-\zeta^2}\,t + \tan^{-1}\frac{\sqrt{1-\zeta^2}}{\zeta}\right]}{\sqrt{1-\zeta^2}}$$

On solving above equation we have expression for rise time equal to

$$t_r = \frac{\pi - \tan^{-1}\dfrac{\sqrt{1-\zeta^2}}{\zeta}}{\omega\sqrt{1-\zeta^2}}$$

Peak Time : On differentiating the expression of $c(t)$ we can obtain the expression for peak time. $dc(t)/dt = 0$ we have expression for peak time,

$$t_p = \frac{\pi}{\omega\sqrt{1-\zeta^2}}$$

Maximum overshoot : Now it is clear from the figure that the maximum overshoot will occur at peak time tp hence on putting the value of peak time we will get maximum overshoot as

$$\%MP = e^{-\zeta\pi/\sqrt{1-\zeta^2}} \times 100$$

Settling Time : Settling time is given by the expression

$$t_s = \frac{4}{\omega\zeta}$$

Steady state error : The steady state error is diffrerence between the actual output and the desired output hence at time tending to infinity the steady state error is zero.

STEADY-STATE ERROR

Steady-state error is defined as the difference between the input (command) and the output of a system in the limit as time goes to infinity (*i.e.* when the response has reached steady state). The steady-state error will depend on the type of input (step, ramp, *etc.*) as well as the system type (0, I, or II).

Note: Steady-state error analysis is only useful for stable systems. You should always check the system for stability before performing a steady-state error analysis. Many of the techniques that we present will give an answer even if the error does not reach a finite steady-state value.

Calculating Steady-state Errors

Before talking about the relationships between steady-state error and system type, we will show how to calculate error regardless of system type or input. Then, we will start deriving formulas we can apply when the system has a specific structure and the input is one of our standard functions. Steady-state error can be calculated from the open- or closed-loop transfer function for unity feedback systems. For example, let's say that we have the system given below.

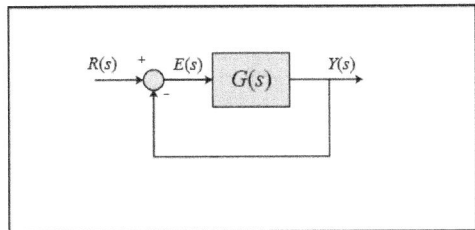

This is equivalent to the following system, where $T(s)$ is the closed-loop transfer function.

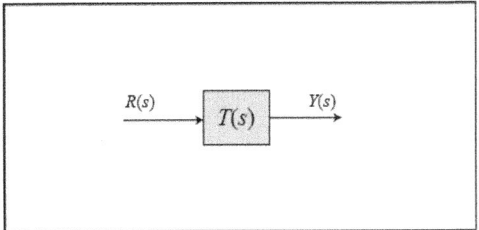

We can calculate the steady-state error for this system from either the open- or closed-loop transfer function using the Final Value Theorem. Recall that this theorem can only be applied if the subject of the limit ($sE(s)$ in this case) has poles with negative real part.

(1) $\quad e(\infty) = \lim\limits_{s \to \infty} sE(s) = \lim\limits_{s \to \infty} \dfrac{sR(s)}{1+G(s)}$

(2) $\quad e(\infty) = \lim\limits_{s \to \infty} sE(s) = \lim\limits_{y \to \infty} sR(s)\left[1-T(s)\right]$

Now, let's plug in the Laplace transforms for some standard inputs and determine equations to calculate steady-state error from the open-loop transfer function in each case.

- Step Input ($R(s) = 1 / s$):

(3) $e(\infty) = \dfrac{1}{1 + \lim_{s \to 0} G(s)} = \dfrac{1}{1 + K_p} \Rightarrow K_p = \lim_{y \to \infty} G(s)$

- Ramp Input ($R(s) = 1 / s^2$):

(4) $e(\infty) = \dfrac{1}{\lim_{s \to 0} sG(s)} = \dfrac{1}{K_v} \Rightarrow K_v = \lim_{s \to 0} sG(s)$

- Parabolic Input ($R(s) = 1 / s^3$):

(5) $e(\infty) = \dfrac{1}{\lim_{s \to 0} s^2 G(s)} = \dfrac{1}{K_a} \Rightarrow K_a = \lim_{s \to 0} s^2 G(s)$

When we design a controller, we usually also want to compensate for disturbances to a system. Let's say that we have a system with a disturbance that enters in the manner shown below.

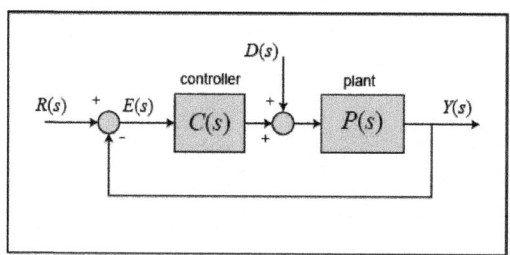

We can find the steady-state error due to a step disturbance input again employing the Final Value Theorem (treat $R(s) = 0$).

(6) $e(\infty) = \lim_{s \to 0} s\,E(s) = \dfrac{1}{\lim_{s \to 0} \frac{1}{P(s)} + \lim_{s \to 0} C(s)}$

When we have a non-unity feedback system we need to be careful since the signal entering $G(s)$ is no longer the actual error $E(s)$. Error is the difference between the commanded reference and the actual output,

$$E(s) = R(s) - Y(s).$$

When there is a transfer function $H(s)$ in the feedback path, the signal being substracted from $R(s)$ is no longer the true output $Y(s)$, it has been distorted by $H(s)$. This situation is depicted below.

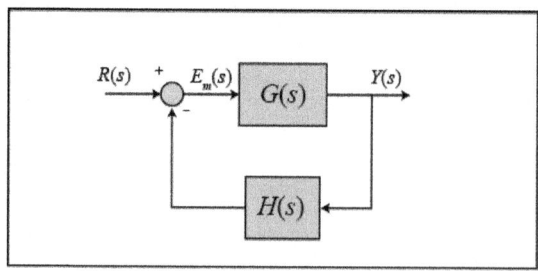

Manipulating the blocks, we can transform the system into an equivalent unity-feedback structure as shown below.

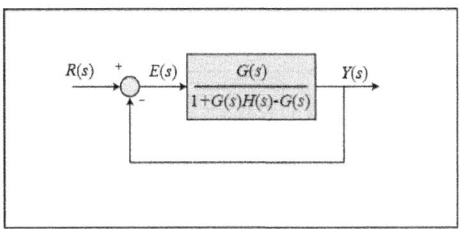

Then we can apply the equations we derived above.

System type and steady-state error

If you refer back to the equations for calculating steady-state errors for unity feedback systems, you will find that we have defined certain constants (known as the static error constants). These constants are the position constant (Kp), the velocity constant (Kv), and the acceleration constant (Ka). Knowing the value of these constants, as well as the system type, we can predict if our system is going to have a finite steady-state error.

First, let's talk about system type. The system type is defined as the number of pure integrators in the forward path of a unity-feedback system. That is, the system type is equal to the value of n when the system is represented as in the following figure. It does not matter if the integrators are part of the controller or the plant.

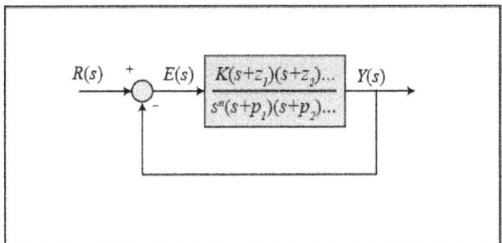

Therefore, a system can be type 0, type 1, *etc.* The following tables summarize how steady-state error varies with system type.

Type 0 system	Step Input	Ramp Input	Parabolic Input
Steady-State Error Formula	1/(1+Kp)	1/Kv	1/Ka
Static Error Constant	Kp = constant	Kv = 0	Ka = 0
Error	1/(1+Kp)	infinity	Infinity

Type 1 system	Step Input	Ramp Input	Parabolic Input
Steady-State Error Formula	1/(1+Kp)	1/Kv	1/Ka
Static Error Constant	Kp = infinity	Kv = constant	Ka = 0
Error	0	1/Kv	Infinity

Type 2 system	Step Input	Ramp Input	Parabolic Input
Steady-State Error Formula	1/(1+Kp)	1/Kv	1/Ka
Static Error Constant	Kp = infinity	Kv = infinity	Ka = constant
Error	0	0	1/Ka

Example: Meeting steady-state error requirements

Consider a system of the form shown below.

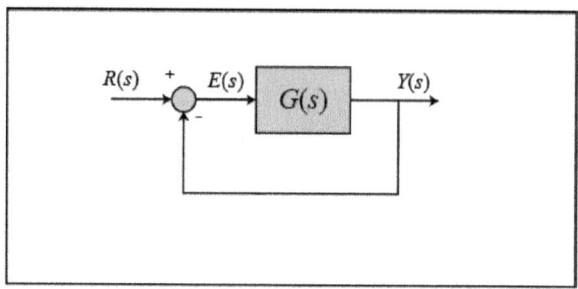

For this example, let $G(s)$ equal the following.

$$(7) \quad G(s) = \frac{K(s+3)(s+5)}{s(s+7)(s+8)}$$

Since this system is type 1, there will be no steady-state error for a step input and there will be infinite error for a parabolic input. The only input that will yield a finite steady-state error in this system is a ramp input. We wish to choose K such that the closed-loop system has a steady-state error of 0.1 in response to a ramp reference. Let's first examine the ramp input response for a gain of $K = 1$.

```
s = tf('s');
G = ((s+3)*(s+5))/(s*(s+7)*(s+8));
T = feedback(G,1);
t = 0:0.1:25;
u = t;
[y,t,x] = lsim(T,u,t);
plot(t,y,'y',t,u,'m')
xlabel('Time (sec)')
ylabel('Amplitude')
title('Input-purple, Output-yellow')
```

The steady-state error for this system is quite large, since we can see that at time 20 seconds the output is approximately 16 as compared to an input of 20 (steady-state error is approximately equal to 4).

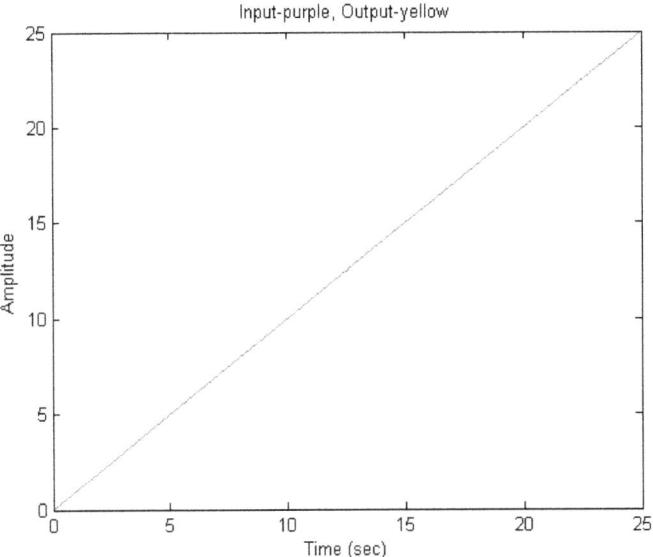

We know from our problem statement that the steady-state error must be 0.1. Therefore, we can solve the problem following these steps:

$$(8) \quad e(\infty) = \frac{1}{K_v} = 0.1$$

$$(9) \quad K_v = 10 = \lim_{s \to 0} sG(s) = \frac{15K}{56}$$

$$(10) \Rightarrow K = 37.33$$

Let's see the ramp input response for $K = 37.33$ by entering the following code in the MATLAB command window.

```
K = 37.33 ;
s = tf('s');
G = (K*(s+3)*(s+5))/(s*(s+7)*(s+8));
sysCL = feedback(G,1);
t = 0:0.1:50;
u = t;
[y,t,x] = lsim(sysCL,u,t);
plot(t,y,'y',t,u,'m')
xlabel('Time (sec)')
ylabel('Amplitude')
title('Input-purple, Output-yellow')
```

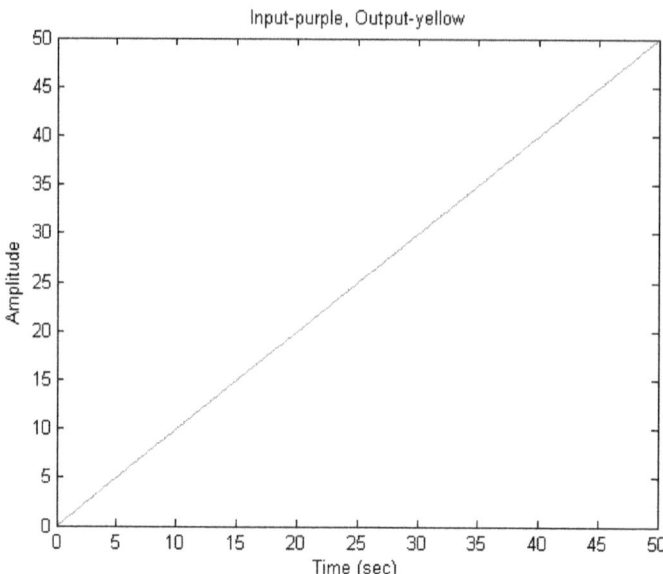

In order to get a better view, we must zoom in on the response. We choose to zoom in between time equals 39.9 and 40.1 seconds because that will ensure that the system has reached steady state.

```
axis([39.9,40.1,39.9,40.1])
```

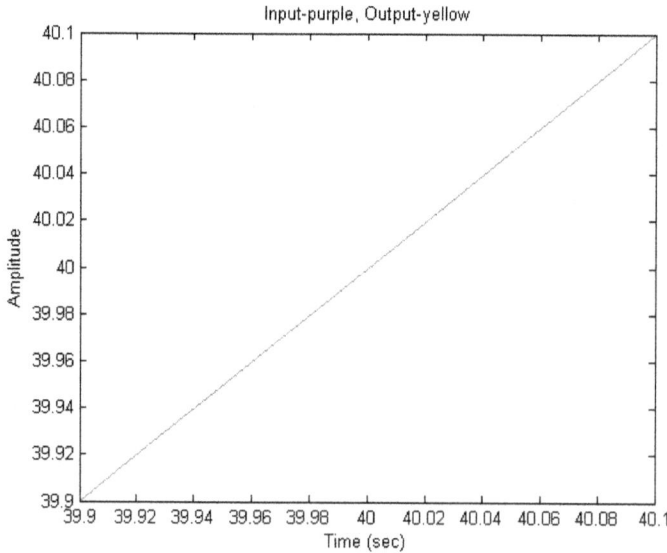

Examination of the above shows that the steady-state error is indeed 0.1 as desired.

Now let's modify the problem a little bit and say that our system has the form shown below.

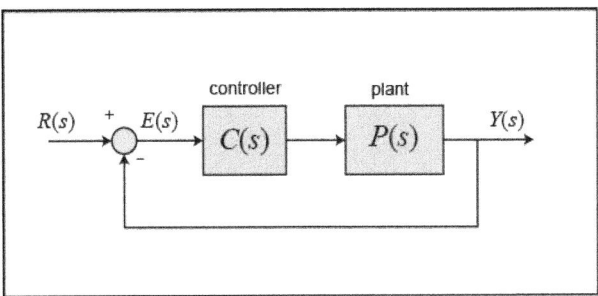

In essence we are no distinguishing between the controller and the plant in our feedback system. Now we want to achieve zero steady-state error for a ramp input.

From our tables, we know that a system of type 2 gives us zero steady-state error for a ramp input. Therefore, we can get zero steady-state error by simply adding an integrator (a pole at the origin). Let's view the ramp input response for a step input if we add an integrator and employ a gain $K = 1$.

```
s = tf('s');
P = ((s+3)*(s+5))/(s*(s+7)*(s+8));
C = 1/s;
sysCL = feedback(C*P,1);
t = 0:0.1:250;
u = t;
[y,t,x] = lsim(sysCL,u,t);
plot(t,y,'y',t,u,'m')
xlabel('Time (sec)')
ylabel('Amplitude')
title('Input-purple, Output-yellow')
```

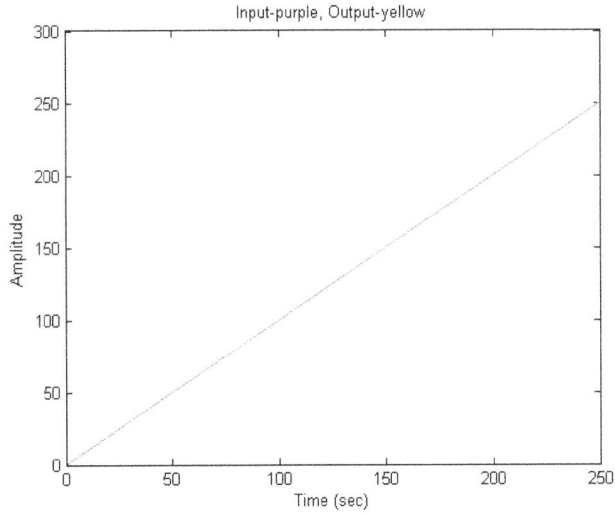

As you can see, there is initially some oscillation (you may need to zoom in). However, at steady state we do have zero steady-state error as desired. Let's zoom in around 240 seconds (trust me, it doesn't reach steady state until then).

```
axis([239.9,240.1,239.9,240.1])
```

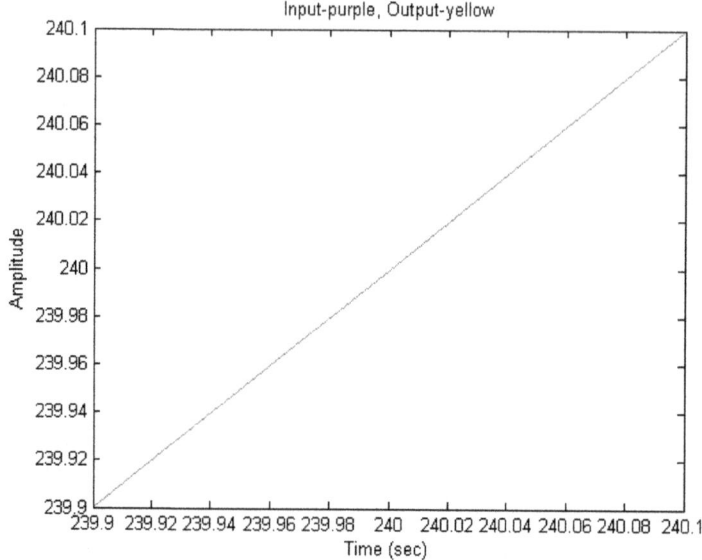

As you can see, the steady-state error is zero. Feel free to zoom in on different areas of the graph to observe how the response approaches steady state.

CONTROLLER (CONTROL THEORY)

In control theory, a controller is a device, historically using mechanical, hydraulic, pneumatic or electronic techniques often in combination, but more recently in the form of a microprocessor or computer, which monitors and physically alters the operating conditions of a given dynamical system. Typical applications of controllers are to hold settings for temperature, pressure, flow or speed.

Input and Control Variables

A system can either be described as a MIMO system, having multiple inputs and outputs, therefore requiring more than one controller; or a SISO system, consisting of a single input and single output, hence having only a single controller. Depending on the set-up of the physical (or non-physical) system, adjusting the system's input variable (assuming it is SISO) will affect the operating parameter, otherwise known as the controlled output variable. Upon receiving the error signal that marks the disparity between the desired value (setpoint) and the actual output value, the controller will then attempt to regulate controlled output behaviour. The controller achieves this by either attenuating or amplifying the input signal to the plant so that the output is returned to the setpoint. For example, a simple feedback control system will generate an error signal that's mathematically depicted as the difference between the setpoint value and the output value, r-y.

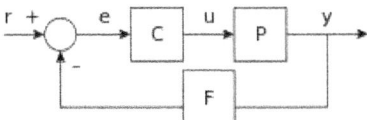

Fig. : A simple feedback control loop illustrates that the error signal is received by controller C, which then either attenuate or amplify the input signal to the plant.

This signal describes the magnitude by which the output value deviates from the setpoint. The signal is subsequently sent to the controller C which then interprets and adjusts for the discrepancy. If the plant is a physical one, the inputs to the system are regulated by means of actuators.

A thermostat on a heater is an example of open loop control that is on or off. A temperature sensor turns the heat source on if the temperature falls below the set point and turns the heat source off when the set point is reached. There is no measurement of the difference between the set point and the measured temperature (*e.g.* no error measurement) and no adjustment to the rate at which heat is added other than all or none.

A familiar example of feedback control is cruise control on an automobile. Here speed is the measured variable. The operator (driver) adjusts the desired speed set point (*e.g.* 100 km/hr) and the controller monitors the speed sensor and compares the measured speed to the set point. Any deviations, such as changes in grade, drag, wind speed or even using a different grade of fuel (for example an ethanol blend) are corrected by the controller making a compensating adjustment to the fuel valve open position, which is the manipulated variable. The controller makes adjustments having information only about the error (magnitude, rate of change or cumulative error) although adjustments known as *tuning* are used to achieve stable control. The operation of such controllers is the subject of control theory.

System	Controlled Outputs Include	Controller	Desired Performance Includes
Aircraft	Course, pitch, roll, yaw	Autopilot	Maintain flight path on a safe and smooth trajectory
Furnace	Temperature	Temperature controller	Follow warm-up temperature profile, then maintain temperature
Water treatment	pH value of effluent	pH controller	Neutralize effluent to specified accuracy
Automobile	Speed	Cruise control	Attain, then maintain selected speed without undue fuel consumption

The notion of controllers can be extended to more complex systems. In the natural world, individual organisms also appear to be equipped with controllers that assure the homeostasis necessary for survival of each individual. Both human-made and natural systems exhibit collective behaviors amongst individuals in which the controllers seek some form of equilibrium

Types of Controlling System

In control theory there are two basic types of control: feedback and feed-forward.

Feedback

The input to a feedback controller is the same as what it is trying to control - the controlled variable is "fed back" into the controller. The thermostat of a house is an example of a feedback controller. This controller relies on measuring the controlled variable, in this case the temperature of the house, and then adjusting the output, whether or not the heater is on. However, feedback control usually results in intermediate periods where the controlled variable is not at the desired set-point. With the thermostat example, if the door of the house were opened on a cold day, the house would cool down. After it fell below the desired temperature (set-point), the heater would kick on, but there would be a period when the house was colder than desired.

Feed-forward

Feed-forward control can avoid the slowness of feedback control. With feed-forward control, the disturbances are measured and accounted for before they have time to affect the system. In the house example, a feed-forward system may measure the fact that the door is opened and automatically turn on the heater before the house can get too cold. The difficulty with feed-forward control is that the effect of the disturbances on the system must be accurately predicted, and there must not be any unmeasured disturbances. For instance, if a window were opened that was not being measured, the feed-forward-controlled thermostat might still let the house cool down.

To achieve the benefits of feedback control (controlling unknown disturbances and not having to know exactly how a system will respond to disturbances) *and* the benefits of feed-forward control (responding to disturbances before they can affect the system), there are combinations of feedback and feed-forward that can be used.

Examples

Some examples of where feedback and feed-forward control can be used together are dead-time compensation, and inverse response compensation. Dead-time compensation is used to control devices that take a long time to show any change to a change in input, for example, change in composition of flow through a long pipe. A dead-time compensation control uses an element (also called a Smith predictor) to predict how changes made now by the controller will affect the controlled variable in the future. The controlled variable is also measured and used in feedback control. Inverse response compensation involves controlling systems where a change at first affects the measured variable one way but later affects it in the opposite way. An example would be eating candy. At first it will give you lots of energy, but later you will be very tired. As can be imagined, it is difficult to control this system with feedback alone, therefore a predictive feed-forward element is necessary to predict the reverse effect that a change will have in the future. "

Types of Controller

Most control valve systems in the past were implemented using mechanical systems or solid state electronics. Pneumatics were often utilized to transmit information and control using pressure. However, most modern industrial control systems now rely on computers for the industrial controller. Obviously it is much easier to implement complex control algorithms on a computer than using a mechanical system.

For feedback controllers there are a few simple types. The most simple is like the thermostat that just turns the heat on if the temperature falls below a certain value and off it exceeds a certain value (on-off control).

Another simple type of controller is a proportional controller. With this type of controller, the controller output (control action) is proportional to the error in the measured variable.

In feedback control, it is standard to define the error as the difference between the desired value (setpoint) y_s and the current value (measured) y. If the error is large, then the control action is large. Mathematically:

$$u(t) = K_c * e(t) + u_0$$

where

$u(t)$ represents the control action (controller output),

$e(t) = y_s(t) - y(t)$ represents the error,

K_c represents the controller's gain, and

u_0 represents the steady state control action (bias) necessary to maintain the variable at the steady state when there is no error.

It is important that the control action u counteracts the change in the controlled variable y (negative feedback). There are then two cases depending on the sign of the process gain.

In the first case the process gain is positive, so an increase in the controlled variable (measurement) y requires a decrease in the control action u (reverse-acting control). In this case the controller gain K_c is positive, because the standard definition of the error already contains a negative sign for y.

In the second case the process gain is negative, so an increase in the controlled variable (measurement) y requires an increase in the control action u (direct-acting control). In this case the controller gain K_c is negative.

A typical example of a reverse-acting system is control of temperature (y) by use of steam (u). In this case the process gain is positive, so if the temperature increases, the steam flow must be decreased to maintain the desired temperature. Conversely, a typical example of a direct-acting system is control of temperature using cooling water. In this case the process gain is negative, so if the temperature increases, the cooling water flow must be increased to maintain the desired temperature.

Although proportional control is simple to understand, it has drawbacks. The largest problem is that for most systems it will never entirely remove error. This is because when error is 0 the controller only provides the steady state control action so the system will settle back to the original steady state (which is probably not the new set point that we want the system to be at). To get the system to operate near the new steady state, the controller gain, K_c, must be very large so the controller will produce the required output when only a very small error is present. Having large gains can lead to system instability or can require physical impossibilities like infinitely large valves.

Alternates to proportional control are proportional-integral (PI) control and proportional-integral-derivative (PID) control. PID control is commonly used to implement closed-loop control.

Open-loop control can be used in systems sufficiently well-characterized as to predict what outputs will necessarily achieve the desired states. For example, the rotational velocity of an electric motor may be well enough characterized for the supplied voltage to make feedback unnecessary.

The drawback of open-loop control is that it requires perfect knowledge of the system (*i.e.* one knows exactly what inputs to give in order to get the desired output), and it assumes there are no disturbances to the system.

TIME RESPONSE OF DISCRETE TIME SYSTEMS

Absolute stability is a basic requirement of all control systems. Apart from that, good relative stability and steady state accuracy are also required in any control system, whether continuous time or discrete time. Transient response corresponds to the system close loop poles and steady state response corresponds to the excitation poles or poles of the input function.

Time Response Specifications

In many practical control systems, the desired performance characteristics are specified in terms of time domain quantities. Unit step input is the most commonly used in analysis purpose of a system since it is easy to generate and represents a sufficiently drastic change thus provides useful information on both transient and steady state response.

The transient response of a system depends on the initial conditions. It is a common practice to consider the system initially at rest.

Consider the digital control system shown in Figure below

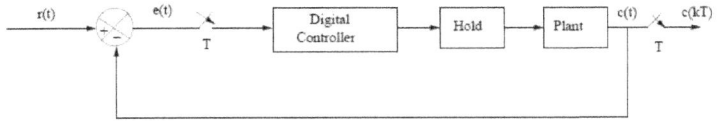

Fig. : Block Diagram of a closed loop digital system

Similar to the continuous time case, transient response of a digital control system can also be characterized by the following :

1. Rise time (t_r): Time required for the unit step response to rise from 0% to 100% of its final value in case of underdamped system or 10% to 90% of its final value in case of overdamped system.

2. 2. Delay time (t_d): Time required for the the unit step response to reach 50\% of its final value.

3. 3. Peak time (t_p): Time at which maximum peak occurs.

4. 4. Peak overshoot (M_p): The difference between the maximum peak and the steady state value of the unit step response.

5. 5. Settling time (t_s): Time required for the unit step response to reach and stay within 2% or 5% of its steady state value.

However since the output response is discrete the calculated performance measures may be slightly different from the actual values. The output has a maximum value c_{max} whereas the maximum value of the discrete output is c^*_{max} which is always less than or equal to c_{max}. If the sampling period is small enough compared to the oscillations of the response then this difference will be small otherwise c^*_{max} may be completely erroneous.

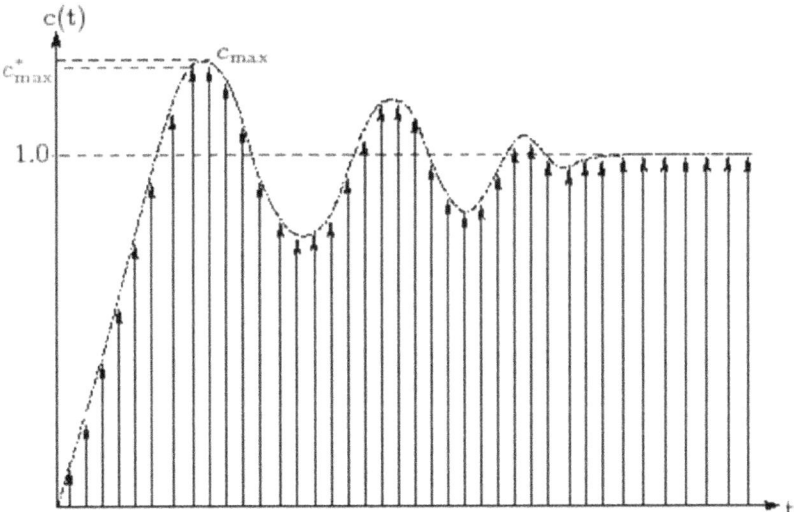

Fig. : Unit step response of a discrete time system.

Steady State Error

The steady state performance of a stable control system is measured by the steady error due to step, ramp or parabolic input depending on the system type. Consider the discrete time system as shown in Figure below.

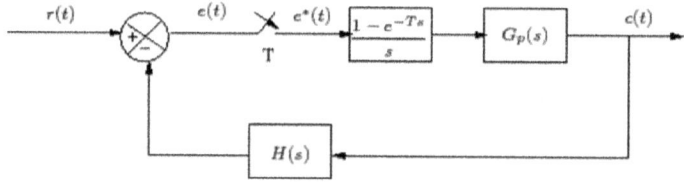

Fig. : Block Diagram 2.

From Figure previous, we can write

$$E(s) = R(s) - H(s)\, C(s)$$

We will consider the steady state error at the sampling instants.

From final value theorem

$$\lim_{k \to \infty} e(kT) = \lim_{z \to \infty} [1 - z^{-1})\, E(z)]$$

$$G(z) = (1 - z^{-1}) Z \left[\frac{G_p(s)}{s} \right]$$

$$GH(z) = (1 - z^{-1}) Z \left[\frac{G_p(s) H(s)}{s} \right]$$

$$\frac{C(z)}{R(z)} = \frac{G(z)}{1 + GH(z)}$$

Again, $E(z) = R(z) - GH(z)\, E(z)$

or, $E(z) = \dfrac{1}{1 + GH(z)} R(z)$

$$\Rightarrow e_{ss} = \lim_{z \to 1} \left[(1 - z^{-1}) \frac{1}{1 + GH(z)} R(z) \right]$$

The steady state error of a system with feedback thus depends on the input signal $R(z)$ and the loop transfer function $GH(z)$.

Type-0 System and Position Error Constant

Systems having a finite nonzero steady state error with a zero order polynomial input (step input) are called **Type-0** systems. The position error constant for a system is defined for a step input.

$r(t) = u_r(t)$ unit ramp

$$R(z) = \frac{Tz}{(z-1)^2} = \frac{TZ^{-1}}{(1-Z^{-1})^2}$$

$$e_{ss} = \lim_{z \to 1} \frac{T}{(z-1)GH(z)} = \frac{1}{K_v}$$

where $K_p = \lim_{z \to 1} GH(z)$ is known as the position error constant.

Type-1 System and Velocity Error Constant

Systems having a finite nonzero steady state error with a first order polynomial input (ramp input) are called **Type-1** systems. The velocity error constant for a system is defined for a ramp input.

$r(t) = u_r(t)$ unit ramp

$$R(z) = \frac{Tz}{(z-1)^2} = \frac{TZ^{-1}}{(1-Z^{-1})^2}$$

$$e_{ss} = \lim_{z \to 1} \frac{T}{(z-1)GH(z)} = \frac{1}{K_v}$$

where $K_v = \frac{1}{T} \lim_{z \to 1} \left[(z-1)GH(z) \right]$ is known as the velocity error constant.

Type-2 System and Acceleration Error Constant

Systems having a finite nonzero steady state error with a second order polynomial input (parabolic input) are called **Type-2** systems. The acceleration error constant for a system is defined for a parabolic input.

$$R(z) = \frac{T^2 z(z+1)}{2(z-1)^3} = \frac{T^2(1+z^{-1})z^{-1}}{2(1-Z^{-1})^3}$$

$$e_{ss} = \frac{T^2}{2} \lim_{z \to 1} \frac{(z+1)}{(z-1)^2 [1+GH(z)]} = \frac{1}{\lim_{z \to 1} \frac{(z-1)^2}{T^2} GH(z)} = \frac{1}{K_a}$$

where $K_a = \lim_{z \to 1} \frac{(z-1)^2}{T^2} GH(z)$ is known as the **acceleration error constant**.

Table shows the steady state errors for different types of systems for different inputs.

Table 1: Steady state errors

System	Step input	Ramp input	Parabolic input
Type-0	$\dfrac{1}{1+kp}$	∞	∞
Type-1	0	$\dfrac{1}{K_v}$	∞
Type-2	0	0	$\dfrac{1}{K_a}$

Example 1: Calculate the steady state errors for unit step, unit ramp and unit parabolic inputs for the system shown in Figure below.

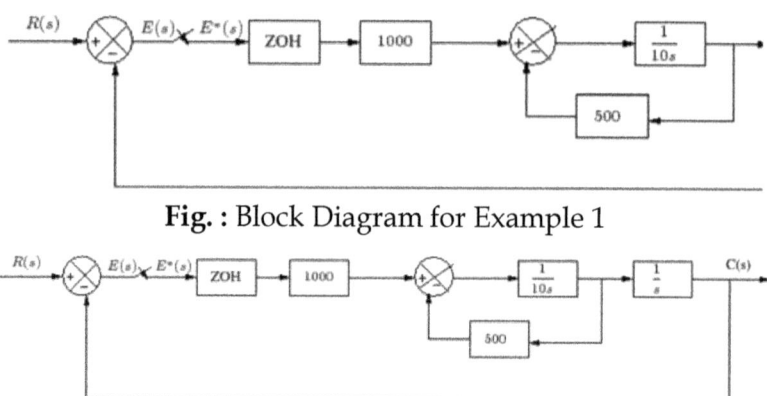

Fig. : Block Diagram for Example 1

PROCESS CONTROL SYSTEMS FOR ENERGY EFFICIENCY

How Process Control Systems Work

Process control systems monitor industrial processes to make sure they don't vary from pre-set limits. They start corrective action if they find unwanted deviations or abnormalities.

Basic Closed-loop Control

The key element of a process control system is the basic closed-loop control. Systems usually include many hundreds of individual control loops, each of which controls one aspect of a particular process - for example, the temperature of an oven. The control loop's job is to make sure that a particular variable or parameter is maintained at its pre-set value, or 'set-point', and doesn't vary from this by more than an acceptable amount.

The control loop must also react quickly to any changes in set-point so that production isn't interrupted.

To achieve this, a control loop is made up of three main components - the measurement device, the controller and the regulator.

Measurement Device

To stop any unwanted variation in a production process, a control loop monitors the process regularly to check it's performing as it should. This is usually done by a sensor that measures a particular property like temperature. A transmitter converts the sensor's output into a signal which is sent to a controller. The signal can be sent individually or with other signals through a special network - called a 'fieldbus'.

The Controller

The controller compares the measurement recorded by the sensor against the pre-set value. If there's an unacceptable difference it initiates appropriate action. For example, if an oven's temperature has fallen too far, it instructs the regulator to send more fuel to the burners to increase it up to the required set-point.

Most of the control tasks required by an industrial process can be handled by a well-designed and well-tuned **single-loop controller**. These normally monitor a single measurement and adjust one regulator, but they can also be linked to another controller to adjust another related set-point. This is known as a 'cascade system'.

The Regulator

The regulator controls the throughput of the process. It responds to commands from the controller and makes adjustments where necessary. Control valves are the most common type of regulator - these adjust the flow of a fluid in response to messages from the controller. A variable speed pump is an alternative type of regulator which controls the flow of a fluid more accurately and with greater energy efficiency. Variable speed drives can also be used where the controller regulates the movement of solid materials rather than fluids or gases.

Upgrading Process Control Systems

If your process control system is more than 20 years old it's worth considering upgrading it. Installing new digital technology can reduce your staffing levels and maintenance costs and improve the efficiency of your production operations. As well as basic closed-loop controls, there are modern systems capable of more complex control.

Sequence Controllers

You can replace single-loop controllers that are used to sequence simple activities, like the mechanical operations involved in process start up and shut down, with sequence controllers such as programmable logic controllers. These can be expanded as different aspects of production are automated. They can include both single-loop and more advanced controllers.

Distributed Control Systems

These are modular systems that control large or complex processes. Operators can adjust the set-points of many different controllers from a central control room. A high-speed network or 'control bus' connects each controller to a central control unit. This sends messages such as fault codes to other high-level systems so that you can make sure you achieve high levels of quality and throughput.

Supervisory Control and Data-acquisition Systems

These sophisticated software packages are used to control a wide range of industrial processes and can store production details so that they can be analysed later. Advanced systems include features that help operators to control and optimise production automatically.

Issues to Consider before Upgrading

If you decide to upgrade your process control system, you should plan carefully as it will involve a high capital cost. You can do this by:

- keeping it simple - ensuring your proposed system isn't unnecessarily complicated and expensive
- carrying out a feasibility study - estimating what the project will cost and the value of the benefits it will bring
- involving team members - making sure everyone who will be managing, operating and maintaining the system is part of the planning
- carrying out a pilot study on part of the process control system
- raising staff awareness about the new system and why it's been installed
- putting operator training in place
- setting up a maintenance schedule to keep the system running efficiently

Maintaining Process Control Systems

The **performance of your process control system** may gradually decline over time because your operators override the system, or your equipment and plant develop faults. You may not be aware that the system is performing badly because maintenance schedules don't always include the right checks to identify problems. If your system is poorly-maintained, your processes won't be operating efficiently and your energy costs will be higher than necessary.

Indications that your system isn't performing well may include:

- operators setting controllers to manual when they should be on automatic mode
- inefficient operation
- too much variability
- frequent calls for repairs and maintenance

The maintenance approach itself may be the cause of many problems. If 'quick fix' breakdown maintenance is carried out in response to equipment failure, underlying problems may never be sorted out and can become permanent. A planned preventive maintenance strategy involves carrying out maintenance tasks regardless of whether a machine has broken down. This can also help you identify opportunities to enhance or upgrade your system.

Produce a Maintenance Manual and Use a Log Book

If you have a good maintenance manual, you will be able to keep process control systems working well and make sure that you don't overlook routine maintenance tasks. You should include in the manual:

- details of all maintenance tasks, who is responsible for them, and how often they must be done
- schematic diagrams of all equipment and controls
- operating instructions including emergency shutdown procedures
- contact details for equipment manufacturers, and installation and maintenance technicians

You should always make sure the key parts of the system are working correctly. Carry out regular checks on:

- instruments - get faulty instruments repaired or recalibrated by qualified personnel
- measurement devices - correct problems caused by poor installation of sensors
- valves - check for 'sticky' valves which increase energy consumption

You should also keep detailed records of all your maintenance activity in a log book. This will help you when:

- dealing with problems that keep happening
- improving process control - for example by training operatives or replacing manual controls with automatic ones
- upgrading your equipment if it has reached the end of its useful life

Fine-tune and Monitor Process Control Systems

Although process control systems only use a small percentage of the total amount of energy consumed in a typical industrial unit, they directly affect how efficiently all the plant and machinery on the site works. So making sure your process control system is well tuned and working correctly is very cost-effective and can lead to substantial energy savings.

The performance of a control system declines over time as the system becomes de-tuned. Tuning it regularly means that it continues to work as well and efficiently as possible. Tuning - or 'optimising' - a process control loop involves adjusting the controller so that it responds more quickly to changes in the process. Tuning requires no capital investment and it is easy to measure the cost-savings from doing so quickly.

Automatic Control

One of the signs that a control system has become de-tuned is when operators frequently override automatic controls and switch to manual operation, which is much less efficient. You should check at key times - such as shift changes - that operators haven't set systems to manual operation unnecessarily.

Staff Training

Making sure your staff are trained to recognise and deal with tuning problems in control systems themselves can reduce the number of calls for maintenance and get processes back to normal quickly.

Monitoring

You should monitor your production processes regularly to help you identify where systems have become inefficient. You should look out for signs that your controls need tuning. These include:

- Excessive or variable energy use. Check that the specific energy consumption per unit of product is consistent. If you measure the energy used per unit of product at different times, but when throughput is the same, this will help determine whether energy use is consistent.
- Over-purification or over-specification of the product. This can occur when the system is using incorrect settings so that it over-reacts to small changes.
- Control disturbances. These can happen if your system isn't set up to compensate automatically for external changes - such as a drop in pressure - before it sets off plant alarms or safety trips.
- Time delays and dead time. Fine tuning the controller can help to avoid dead time caused by the slow response of the system to changes.

As well as highlighting where systems need attention, monitoring also helps you to establish where you could put in place energy-saving measures in the future.

Top Tips for Improving Process Control

Following the steps below will help you **improve the way your process control system operates** and reduce your energy costs.

Investigate the Current System

Monitor your system's performance and walk around your site looking for signs of poor control. These include inconsistencies in product quality, variations in the energy used per unit of product, controls set to manual instead of automatic, and production upsets.

Check Instruments and Regulators

Ensure that all your existing instruments and regulators have been correctly installed and are working well. You should repair any faulty equipment, fix 'sticky' valves and check that measurement devices like sensors and probes have been properly installed, replacing them if necessary.

Make Sure Controls are Working Well

Check that control loops are not always set to manual and make sure operators are aware that working in automatic mode is much more efficient. Review control loops to check they're correctly configured and tuned and adjust controllers to minimise time delays and dead time.

Identify Improvements in Control

Draw up a list to show where you could save energy by improving your systems control. You should prioritise low-cost actions and enhancements and think about when to start larger projects. Identify the capital cost and payback period for each and put together an action plan for implementation.

Plan for Implementation

Put together the business case for the implementation of your plan. This should include a feasibility study of costs and savings, a detailed risks/benefits assessment and information about the project and the project team. A pilot study could help decision-makers to give support to the project.

Take Action and Improve Controls

Ensure that all new control equipment is installed and calibrated properly. You should train operators to use the new systems and how to deal with unexpected performance after the changes have been made. You should also set new energy consumption targets and draw up a preventive maintenance schedule to keep controls running well and run a campaign to increase awareness of the new systems and changes that have been made among your staff.

Monitor Performance

Set up a regular system of checks to identify any areas of poor control as early as possible, perhaps when investigating your current system. This should include reviewing operator logs for symptoms of poor control, setting performance targets and comparing them with independent benchmarks, and reviewing targets regularly to make sure they're appropriate.

FEEDBACK SYSTEMS

Feedback Systems process signals and as such are signal processors. The processing part of a feedback system may be electrical or electronic, ranging from a very simple to a highly complex circuits. Simple analogue feedback control circuits can be constructed using individual or discrete components, such as transistors, resistors and capacitors, etc, or by using microprocessor-based and integrated circuits (IC's) to form more complex digital feedback systems.

As we have seen, open-loop systems are just that, open ended, and no attempt is made to compensate for changes in circuit conditions or changes in load conditions due to variations in circuit parameters, such as gain and stability, temperature, supply voltage variations and/or external disturbances. But the effects of these "open-loop" variations can be eliminated or at least considerably reduced by the introduction of Feedback.

A feedback system is one in which the output signal is sampled and then fed back to the input to form an error signal that drives the system. In the previous tutorial about Closed-loop Systems, we saw that in general, Feedback is comprised of a subcircuit that allows a fraction of the output signal from a system to modify the effective input signal in such a way as to produce a response that can differ substantially from the response produced in the absence of such feedback.

Feedback Systems are very useful and widely used in amplifier circuits, oscillators, process control systems as well as other types of electronic systems. But for feedback to be an effective tool it must be controlled as an uncontrolled system will either oscillate or fail to function. The basic model of a feedback system is given as:

Feedback System Block Diagram Model

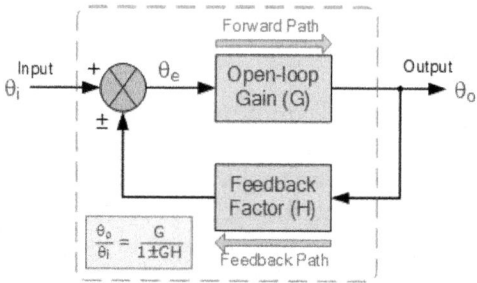

This basic feedback loop of sensing, controlling and actuation is the main concept behind a feedback control system and there are several good reasons why feedback is applied and used in electronic circuits:

- Circuit characteristics such as the systems gain and response can be precisely controlled.
- Circuit characteristics can be made independent of operating conditions such as supply voltages or temperature variations.
- Signal distortion due to the non-linear nature of the components used can be greatly reduced.
- The Frequency Response, Gain and Bandwidth of a circuit or system can be easily controlled to within tight limits.

Whilst there are many different types of control systems, there are just two main types of feedback control namely: Negative Feedback and Positive Feedback.

Positive Feedback Systems

In a "positive feedback control system", the set point and output values are added together by the controller as the feedback is "in-phase" with the input. The effect of positive (or regenerative) feedback is to "increase" the systems gain, *i.e.*, the overall gain with positive feedback applied will be greater than the gain without feedback. For example, if someone praises you or gives you positive feedback about something, you feel happy about yourself and are full of energy, you feel more positive.

However, in electronic and control systems to much praise and positive feedback can increase the systems gain far too much which would give rise to oscillatory circuit responses as it increases the magnitude of the effective input signal.

An example of a positive feedback systems could be an electronic amplifier based on an operational amplifier, or op-amp as shown.

Positive Feedback System

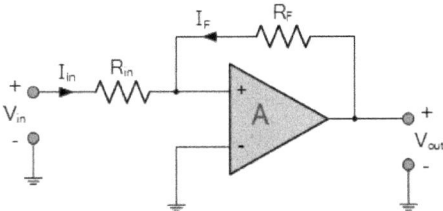

Positive feedback control of the op-amp is achieved by applying a small part of the output voltage signal at V_{out} back to the non-inverting (+) input terminal *via* the feedback resistor, R_F.

If the input voltage V_{in} is positive, the op-amp amplifies this positive signal and the output becomes more positive. Some of this output voltage is returned back to the input by the feedback network.

Thus the input voltage becomes more positive, causing an even larger output voltage and so on. Eventually the output becomes saturated at its positive supply rail.

Likewise, if the input voltage V_{in} is negative, the reverse happens and the op-amp saturates at its negative supply rail. Then we can see that positive feedback does not allow the circuit to function as an amplifier as the output voltage quickly saturates to one supply rail or the other, because with positive feedback loops "more leads to more" and "less leads to less".

Then if the loop gain is positive for any system the transfer function will be:

$$Av = G / (1 - GH).$$

Note that if GH = 1 the system gain Av = infinity and the circuit will start to self-oscillate, after which no input signal is needed to maintain oscillations, which is useful if you want to make an oscillator.

Although often considered undesirable, this behaviour is used in electronics to obtain a very fast switching response to a condition or signal. One example of the use of positive feedback is hysteresis in which a logic device or system maintains a given state until some input crosses a preset threshold. This type of behaviour is called "bi-stability" and is often associated with logic gates and digital switching devices such as multivibrators.

We have seen that positive or regenerative feedback increases the gain and the possibility of instability in a system which may lead to self-oscillation and as such, positive feedback is widely used in oscillatory circuits such as Oscillators and Timing circuits.

Negative Feedback Systems

In a "negative feedback control system", the set point and output values are subtracted from each other as the feedback is "out-of-phase" with the original input. The effect of negative (or degenerative) feedback is to "reduce" the gain. For example, if someone criticises you or gives you negative feedback about something, you feel unhappy about yourself and therefore lack energy, you feel less positive.

Because negative feedback produces stable circuit responses, improves stability and increases the operating bandwidth of a given system, the majority of all control and feedback systems is degenerative reducing the effects of the gain.

An example of a negative feedback system is an electronic amplifier based on an operational amplifier as shown.

Negative Feedback System

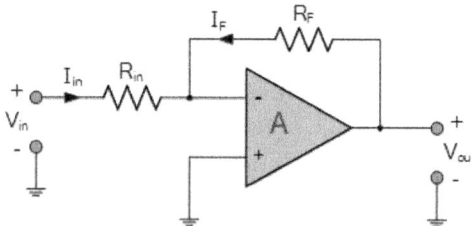

Negative feedback control of the amplifier is achieved by applying a small part of the output voltage signal at V_{out} back to the inverting (–) input terminal *via* the feedback resistor, R_f.

If the input voltage V_{in} is positive, the op-amp amplifies this positive signal, but because its connected to the inverting input of the amplifier, and the output becomes more negative. Some of this output voltage is returned back to the input by the feedback network of R_f.

Thus the input voltage is reduced by the negative feedback signal, causing an even smaller output voltage and so on. Eventually the output will settle down and become stabilised at a value determined by the gain ratio of $R_f \div R_{in}$.

Likewise, if the input voltage V_{in} is negative, the reverse happens and the op-amps output becomes positive (inverted) which adds to the negative input signal. Then we can see that negative feedback allows the circuit to function as an amplifier, so long as the output is within the saturation limits.

So we can see that the output voltage is stabilised and controlled by the feedback, because with negative feedback loops "more leads to less" and "less leads to more".

Then if the loop gain is positive for any system the transfer function will be:

$$Av = G / (1 + GH).$$

The use of negative feedback in amplifier and process control systems is widespread because as a rule negative feedback systems are more stable than positive feedback systems, and a negative feedback system is said to be stable if it does not oscillate by itself at any frequency except for a given circuit condition.

Another advantage is that negative feedback also makes control systems more immune to random variations in component values and inputs. Of course nothing is for free, so it must be used with caution as negative feedback significantly modifies the operating characteristics of a given system.

Classification of Feedback Systems

Thus far we have seen the way in which the output signal is "fed back" to the input terminal, and for feedback systems this can be of either, Positive Feedback or Negative Feedback. But the manner in which the output signal is measured and introduced into the input circuit can be very different leading to four basic classifications of feedback.

Based on the input quantity being amplified, and on the desired output condition, the input and output variables can be modeled as either a voltage or a current. As a result, there are four basic classifications of single-loop feedback system in which the output signal is fed back to the input and these are:

- Series-Shunt Configuration – Voltage in and Voltage out or Voltage Controlled Voltage Source (VCVS).
- Shunt-Shunt Configuration – Current in and Voltage out or Current Controlled Voltage Source (CCVS).
- Series-Series Configuration – Voltage in and Current out or Voltage Controlled Current Source (VCCS).
- Shunt-Series Configuration – Current in and Current out or Current Controlled Current Source (CCCS).

These names come from the way that the feedback network connects between the input and output stages as shown.

Series-Shunt Feedback Systems

Series-Shunt Feedback, also known as *series voltage feedback*, operates as a voltage-voltage controlled feedback system. The error voltage fed back from the feedback network is in *series* with the input. The voltage which is fed back from the output being proportional to the output voltage, V_o as it is parallel, or shunt connected.

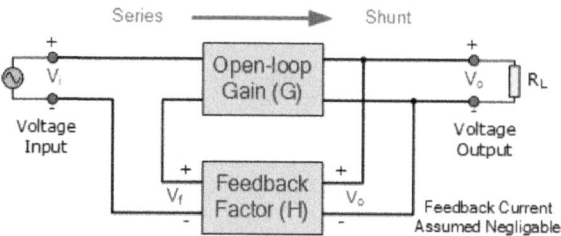

Fig. : Series-Shunt Feedback System.

For the series-shunt connection, the configuration is defined as the output voltage to the input voltage. Most inverting and non-inverting operational amplifier circuits operate with series-shunt feedback producing what is known as a "voltage amplifier". As a voltage amplifier the ideal input resistance, R_{in} is very large, and the ideal output resistance, R_{out} is very small.

Then the "series-shunt feedback configuration" works as a true voltage amplifier as the input signal is a voltage and the output signal is a voltage, so the transfer gain is given as:

$$A_v = V_{out} \div V_{in}.$$

Shunt-Series Feedback Systems

Shunt-Series Feedback, also known as *shunt current feedback*, operates as a current-current controlled feedback system. The feedback signal is proportional to the output current, Io flowing in the load. The feedback signal is fed back in parallel or *shunt* with the input as shown.

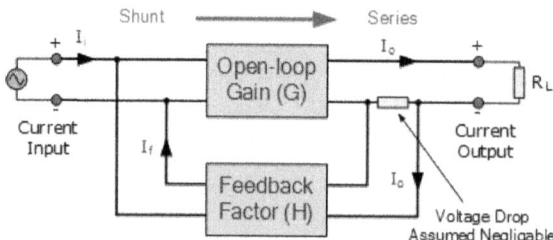

Fig. : Shunt-Series Feedback System.

For the shunt-series connection, the configuration is defined as the output current to the input current. In the shunt-series feedback configuration the signal fed back is in parallel with the input signal and as such its the currents, not the voltages that add.

This parallel shunt feedback connection will not normally affect the voltage gain of the system, since for a voltage output a voltage input is required. Also, the series connection at the output increases output resistance, Rout while the shunt connection at the input decreases the input resistance, R_{in}.

Then the "shunt-series feedback configuration" works as a true current amplifier as the input signal is a current and the output signal is a current, so the transfer gain is given as:

$$A_i = I_{out} \div I_{in}.$$

Series-Series Feedback Systems

Series-Series Feedback Systems, also known as *series current feedback*, operates as a voltage-current controlled feedback system. In the series current configuration the feedback error signal is in *series* with the input and is proportional to the load current, I_{out}. Actually, this type of feedback converts the current signal into a voltage which is actually fed back and it is this voltage which is subtracted from the input.

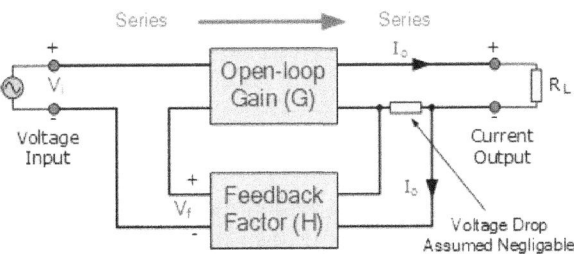

Fig. : Series-Series Feedback System.

For the series-series connection, the configuration is defined as the output current to the input voltage. Because the output current, Io of the series connection is fed back as a voltage, this increases both the input and output impedances of the system. Therefore, the circuit works best as a transconductance amplifier with the ideal input resistance, R_{in} being very large, and the ideal output resistance, Rout is also very large.

Then the "series-series feedback configuration" functions as trans conductance type amplifier system as the input signal is a voltage and the output signal is a current. Then for a series-series feedback circuit the transfer gain is given as:

$$G_m = V_{out} \div I_{in}.$$

Shunt-Shunt Feedback Systems

Shunt-Shunt Feedback Systems, also known as *shunt voltage feedback*, operates as a current-voltage controlled feedback system. In the shunt-shunt feedback configuration the signal fed back is in parallel with the input signal. The output voltage is sensed and the current is subtracted from the input current in shunt, and as such its the currents, not the voltages that subtract.

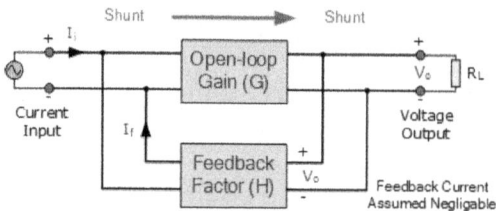

Fig. : Shunt-Shunt Feedback System.

For the shunt-shunt connection, the configuration is defined as the output voltage to the input current. As the output voltage is fed back as a current to a current-driven input port, the shunt connections at both the input and output terminals reduce the input and output impedance therefore the system works best as a trans resistance system with the ideal input resistance, R_{in} being very small, and the ideal output resistance, Rout also being very small.

Then the shunt voltage configuration works as trans resistance type voltage amplifier as the input signal is a current and the output signal is a voltage, so the transfer gain is given as:

$$R_m = I_{out} \div V_{in}.$$

Feedback Systems Summary

We have seen that a Feedback System is one in which the output signal is sampled and then fed back to the input to form an error signal that drives the system, and depending on the type of feedback used, the feedback signal which is mixed with the systems input signal, can be either a voltage or a current.

Feedback will always change the performance of a system and feedback arrangements can be either positive (regenerative) or negative (degenerative) type feedback systems. If the feedback loop around the system produces a loop-gain which is negative, the feedback is said to be negative or degenerative with the main effect of the negative feedback is in reducing the systems gain.

If however the gain around the loop is positive, the system is said to have positive feedback or regenerative feedback. The effect of positive feedback is to increase the gain which can cause a system to become unstable and oscillate especially if GH = -1.

We have also seen that block-diagrams can be used to demonstrate the various types of feedback systems. In the block diagrams above, the input and output variables can be modeled as either a voltage or a current and as such there are four combinations of inputs and outputs that represent the possible types of feedback, namely: Series Voltage Feedback, Shunt Voltage Feedback, Series Current Feedback and Shunt Current Feedback.

The names for these different types of feedback systems are derived from the way that the feedback network connects between the input and output stages either in parallel (shunt) or series.

CONTROL THEORY

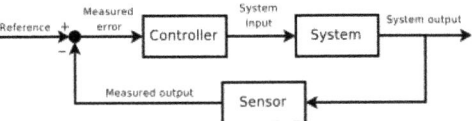

Fig. : The concept of the feedback loop to control the dynamic behavior of the system: this is negative feedback, because the sensed value is subtracted from the desired value to create the error signal, which is amplified by the controller.

Control theory is an interdisciplinary branch of engineering and mathematics that deals with the behavior of dynamical systems with inputs, and how their behavior is modified by feedback. The usual objective of control theory is to control a system, often called the *plant*, so its output follows a desired control signal, called the *reference*, which may be a fixed or changing value. To do this a *controller* is designed, which monitors the output and compares it with the reference. The difference between actual and desired output, called the *error* signal, is applied as feedback to the input of the system, to bring the actual output closer to the reference. Some topics studied in control theory are stability (whether the output will converge to the reference value or oscillate about it), controllability and observability.

Extensive use is usually made of a diagrammatic style known as the block diagram. The transfer function, also known as the system function or network function, is a mathematical representation of the relation between the input and output based on the differential equations describing the system.

Although a major application of control theory is in control systems engineering, which deals with the design of process control systems for industry, other applications range far beyond this. As the general theory of feedback systems, control theory is useful wherever feedback occurs. A few examples are in physiology, electronics, climate modeling, machine design, ecosystems, navigation, neural networks, predator-prey interaction, gene expression, and production theory.

Overview

Control theory is:

- a theory that deals with influencing the behavior of dynamical systems
- an interdisciplinary subfield of science, which originated in engineering and mathematics, and evolved into use by the social sciences, such as economics, psychology, sociology, criminology and in the financial system.

Control systems may be thought of as having four functions: measure, compare, compute and correct. These four functions are completed by five elements: detector, transducer, transmitter, controller and final control element. The measuring function is completed by the detector, transducer and transmitter.

In practical applications these three elements are typically contained in one unit. A standard example of a measuring unit is a resistance thermometer. The compare and compute functions are completed within the controller, which may be implemented electronically by proportional control, a PI controller, PID controller, bistable, hysteretic control or programmable logic controller. Older controller units have been mechanical, as in a centrifugal governor or a carburetor. The correct function is completed with a final control element. The final control element changes an input or output in the control system that affects the manipulated or controlled variable.

An Example

An example of a control system is a car's cruise control, which is a device designed to maintain vehicle speed at a constant *desired* or *reference* speed provided by the driver. The *controller* is the cruise control, the *plant* is the car, and the *system* is the car and the cruise control. The system output is the car's speed, and the control itself is the engine's throttle position which determines how much power the engine delivers.

A primitive way to implement cruise control is simply to lock the throttle position when the driver engages cruise control. However, if the cruise control is engaged on a stretch of flat road, then the car will travel slower going uphill and faster when going downhill. This type of controller is called an *open-loop controller* because there is no feedback; no measurement of the system output (the car's speed) is used to alter the control (the throttle position.) As a result, the controller cannot compensate for changes acting on the car, like a change in the slope of the road.

In a *closed-loop control system*, data from a sensor monitoring the car's speed (the system output) enters a controller which continuously subtracts the quantity representing the speed from the reference quantity representing the desired speed. The difference, called the error, determines the throttle position (the control). The result is to match the car's speed to the reference speed (maintain the desired system output). Now, when the car goes uphill, the difference between the input (the sensed speed) and the reference continuously determines the throttle position. As the sensed speed drops below the reference, the difference increases, the throttle opens, and engine power increases, speeding up the vehicle. In this way, the controller dynamically counteracts changes to the car's speed. The central idea of these control systems is the *feedback loop*, the controller affects the system output, which in turn is measured and fed back to the controller.

Classification

Linear versus Nonlinear Control Theory

The field of control theory can be divided into two branches:

- *Linear control theory* - This applies to systems made of devices which obey the superposition principle, which means roughly that the output is proportional to the input. They are governed by linear differential equations. A major subclass is systems which in addition have parameters which do not change with time, called linear time invariant (LTI) systems. These systems are amenable to powerful frequency domain mathematical techniques of great generality, such as the Laplace transform, Fourier transform, Z transform, Bode plot, root locus, and Nyquist stability criterion. These lead to a description of the system using terms like bandwidth, frequency response, eigenvalues, gain, resonant frequencies, poles, and zeros, which give solutions for system response and design techniques for most systems of interest.

- *Nonlinear control theory* - This covers a wider class of systems that do not obey the superposition principle, and applies to more real-world systems, because all real control systems are nonlinear. These systems are often governed by nonlinear differential equations. The few mathematical techniques which have been developed to handle them are more difficult and much less general, often applying only to narrow categories of systems. These include limit cycle theory, Poincaré maps, Lyapunov stability theorem, and describing functions. Nonlinear systems are often analyzed using numerical methods on computers, for example by simulating their operation using a simulation language. If only solutions near a stable point are of interest, nonlinear systems can often be linearized by approximating them by a linear system using perturbation theory, and linear techniques can be used.

Frequency Domain versus Time Domain

Mathematical techniques for analyzing and designing control systems fall into two different categories:

- *Frequency domain* - In this type the values of the state variables, the mathematical variables representing the system's input, output and feedback are represented as functions of frequency. The input signal and the system's transfer function are converted from time functions to functions of frequency by a transform such as the Fourier transform, Laplace transform, or Z transform. The advantage of this technique is that it results in a simplification of the mathematics; the differential equations that represent the system are replaced by algebraic equations in the frequency domain which are much simpler to solve. However, frequency domain techniques can only be used with linear systems, as mentioned above.

- *Time-domain state space representation* - In this type the values of the state variables are represented as functions of time. With this model the system being analyzed is represented by one or more differential equations. Since frequency domain techniques are limited to linear systems, time domain is widely used to analyze real-world nonlinear systems. Although these are more difficult to solve, modern computer simulation techniques such as simulation languages have made their analysis routine.

SISO vs MIMO

Control systems can be divided into different categories depending on the number of inputs and outputs.

- *Single-input single-output (SISO)* - This is the simplest and most common type, in which one output is controlled by one control signal. Examples are the cruise control example above, or an audio system, in which the control input is the input audio signal and the output is the sound waves from the speaker.

- *Multiple-input multiple-output (MIMO)* - These are found in more complicated systems. For example, modern large telescopes such as the Keck and MMT have mirrors composed of many separate segments each controlled by an actuator. The shape of the entire mirror is constantly adjusted by a MIMO active optics control system using input from multiple sensors at the focal plane, to compensate for changes in the mirror shape due to thermal expansion, contraction, stresses as it is rotated and distortion of the wavefront due to turbulence in the atmosphere. Complicated systems such as nuclear reactors and human cells are simulated by computer as large MIMO control systems.

History

Fig. : Centrifugal governor in a Boulton & Watt engine of 1788.

Although control systems of various types date back to antiquity, a more formal analysis of the field began with a dynamics analysis of thecentrifugal governor, conducted by the physicist James Clerk Maxwell in 1868, entitled *On Governors*. This described and analyzed the phenomenon of self-oscillation, in which lags in the system may lead to overcompensation and unstable behavior.

This generated a flurry of interest in the topic, during which Maxwell's classmate, Edward John Routh, abstracted Maxwell's results for the general class of linear systems. Independently, Adolf Hurwitz analyzed system stability using differential equations in 1877, resulting in what is now known as the Routh–Hurwitz theorem.

A notable application of dynamic control was in the area of manned flight. The Wright brothers made their first successful test flights on December 17, 1903 and were distinguished by their ability to control their flights for substantial periods (more so than the ability to produce lift from an airfoil, which was known). Continuous, reliable control of the airplane was necessary for flights lasting longer than a few seconds.

By World War II, control theory was an important part of fire-control systems, guidance systems and electronics.

Sometimes, mechanical methods are used to improve the stability of systems. For example, ship stabilizers are fins mounted beneath the waterline and emerging laterally. In contemporary vessels, they may be gyroscopically controlled active fins, which have the capacity to change their angle of attack to counteract roll caused by wind or waves acting on the ship.

The Sidewinder missile uses small control surfaces placed at the rear of the missile with spinning disks on their outer surfaces and these are known as rollerons. Airflow over the disks spins them to a high speed. If the missile starts to roll, the gyroscopic force of the disks drives the control surface into the airflow, cancelling the motion. Thus, the Sidewinder team replaced a potentially complex control system with a simple mechanical solution.

The Space Race also depended on accurate spacecraft control, and control theory has also seen an increasing use in fields such as economics.

People in Systems and Control

Many active and historical figures made significant contribution to control theory including:

- Pierre-Simon Laplace (1749-1827) invented the Z-transform in his work on probability theory, now used to solve discrete-time control theory problems. The Z-transform is a discrete-time equivalent of the Laplace transform which is named after him.
- Alexander Lyapunov (1857–1918) in the 1890s marks the beginning of stability theory.
- Harold S. Black (1898–1983), invented the concept of negative feedback amplifiers in 1927. He managed to develop stable negative feedback amplifiers in the 1930s.
- Harry Nyquist (1889–1976) developed the Nyquist stability criterion for feedback systems in the 1930s.

- Richard Bellman (1920–1984) developed dynamic programming since the 1940s.
- Andrey Kolmogorov (1903–1987) co-developed the Wiener–Kolmogorov filter in 1941.
- Norbert Wiener (1894–1964) co-developed the Wiener–Kolmogorov filter and coined the term cybernetics in the 1940s.
- John R. Ragazzini (1912–1988) introduced digital control and the use of Z-transform in control theory (invented by Laplace) in the 1950s.
- Lev Pontryagin (1908–1988) introduced the maximum principle and the bang-bang principle.
- Pierre-Louis Lions (1956) developed viscosity solutions into stochastic control and optimal control methods.

Classical Control Theory

To overcome the limitations of the open-loop controller, control theory introduces feedback. A closed-loop controller uses feedback to control states or outputs of a dynamical system. Its name comes from the information path in the system: process inputs (*e.g.*, voltage applied to an electric motor) have an effect on the process outputs (*e.g.*, speed or torque of the motor), which is measured with sensors and processed by the controller; the result (the control signal) is "fed back" as input to the process, closing the loop.

Closed-loop controllers have the following advantages over open-loop controllers:

- disturbance rejection (such as hills in the cruise control example above)
- guaranteed performance even with model uncertainties, when the model structure does not match perfectly the real process and the model parameters are not exact
- unstable processes can be stabilized
- reduced sensitivity to parameter variations
- improved reference tracking performance

In some systems, closed-loop and open-loop control are used simultaneously. In such systems, the open-loop control is termed feedforward and serves to further improve reference tracking performance.

A common closed-loop controller architecture is the PID controller.

Closed-loop Transfer Function

The output of the system $y(t)$ is fed back through a sensor measurement F to the reference value $r(t)$. The controller C then takes the error e (difference) between the reference and the output to change the inputs u to the system under control P. This is shown in the figure. This kind of controller is a closed-loop controller or feedback controller.

This is called a single-input-single-output (*SISO*) control system; *MIMO* (*i.e.,* Multi-Input-Multi-Output) systems, with more than one input/output, are common. In such cases variables are represented through vectors instead of simple scalar values. For some distributed parameter systems the vectors may be infinite-dimensional (typically functions).

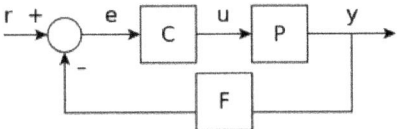

If we assume the controller C, the plant P, and the sensor F are linear and time-invariant (*i.e.,* elements of their transfer function C(s), P(s), and F(s) do not depend on time), the systems above can be analysed using the Laplace transform on the variables. This gives the following relations:

$Y(s) = P(s)\, U(s)$

$U(s) = C(s)\, E(s)$

$E(s) = R(s) - F(s)\, Y(s).$

Solving for Y(s) in terms of R(s) gives:

$$Y(s) = \left(\frac{P(s)\,C(s)}{1 + F(s)\,P(s)\,C(s)} \right) R(s) = H(s)\,R(s).$$

The expression $H(s) = \dfrac{P(s)\,C(s)}{1 + F(s)\,P(s)\,C(s)}$ is referred to as the *closed-loop*

transfer function of the system. The numerator is the forward (open-loop) gain from r to y, and the denominator is one plus the gain in going around the feedback loop, the so-called loop gain. If $|\,P(s)\,C(s)\,| >> 1$, *i.e.,* it has a large norm with each value of s, and if $|\,F(s)\,| \approx 1$, then Y(s) is approximately equal to R(s) and the output closely tracks the reference input.

PID Controller

The PID controller is probably the most-used feedback control design. *PID* is an initialism for *Proportional-Integral-Derivative*, referring to the three terms operating on the error signal to produce a control signal. If u(t) is the control signal sent to the system, y(t) is the measured output and r(t) is the desired output, and tracking error e(t) = r(t) − y(t), a PID controller has the general form

$$u(t) = K_p e(t) + K_I \int e(t)\,dt + K_D \frac{d}{dt} e(t).$$

The desired closed loop dynamics is obtained by adjusting the three parameters K_p, K_I and K_D, often iteratively by "tuning" and without specific knowledge of a plant model. Stability can often be ensured using only the proportional term.

The integral term permits the rejection of a step disturbance (often a striking specification in process control). The derivative term is used to provide damping or shaping of the response. PID controllers are the most well established class of control systems: however, they cannot be used in several more complicated cases, especially if MIMO systems are considered.

Applying Laplace transformation results in the transformed PID controller equation

$$u(s) = K_p e(s) + K_I \frac{1}{s} e(s) + K_D s e(s)$$

$$u(s) = \left(K_p + K_I \frac{1}{s} + K_D s \right) e(s)$$

with the PID controller transfer function

$$C(s) = \left(K_p + K_I \frac{1}{s} + K_D s \right).$$

There exists a nice example of the closed-loop system discussed above. If we take:

PID controller transfer function in series form

$$C(s) = K \left(1 + \frac{1}{sT_i} \right) (1 + sT_d)$$

1st order filter in feedback loop

$$F(s) = \frac{1}{1 + sT_f}$$

linear actuator with filtered input

$$P(s) = \frac{A}{1 + sT_p}, A = \text{const}$$

and insert all this into expression for closed-loop transfer function H(s), then tuning is very easy: simply put

$$K = \frac{1}{A}, T_i = T_f, T_d = T_p$$

and get H(s) = 1 identically.

For practical PID controllers, a pure differentiator is neither physically realisable nor desirable due to amplification of noise and resonant modes in the system. Therefore, a phase-lead compensator type approach is used instead, or a differentiator with low-pass roll-off.

Modern Control Theory

In contrast to the frequency domain analysis of the classical control theory, modern control theory utilizes the time-domain state space representation, a mathematical model of a physical system as a set of input, output and state variables related by first-order differential equations. To abstract from the number of inputs, outputs and states, the variables are expressed as vectors and the differential and algebraic equations are written in matrix form (the latter only being possible when the dynamical system is linear). The state space representation (also known as the "time-domain approach") provides a convenient and compact way to model and analyze systems with multiple inputs and outputs. With inputs and outputs, we would otherwise have to write down Laplace transforms to encode all the information about a system. Unlike the frequency domain approach, the use of the state-space representation is not limited to systems with linear components and zero initial conditions. "State space" refers to the space whose axes are the state variables. The state of the system can be represented as a vector within that space.

Topics in Control Theory

Stability

The *stability* of a general dynamical system with no input can be described with Lyapunov stability criteria.

- A linear system is called bounded-input bounded-output (BIBO) stable if its output will stay bounded for any bounded input.
- Stability for nonlinear systems that take an input is input-to-state stability (ISS), which combines Lyapunov stability and a notion similar to BIBO stability.

For simplicity, the following descriptions focus on continuous-time and discrete-time linear systems.

Mathematically, this means that for a causal linear system to be stable all of the poles of its transfer function must have negative-real values, *i.e.* the real part of each pole must be less than zero. Practically speaking, stability requires that the transfer function complex poles reside:

- in the open left half of the complex plane for continuous time, when the Laplace transform is used to obtain the transfer function.
- inside the unit circle for discrete time, when the Z-transform is used.

The difference between the two cases is simply due to the traditional method of plotting continuous time versus discrete time transfer functions. The continuous Laplace transform is in Cartesian coordinates where the x axis is the real axis and the discrete Z-transform is in circular coordinates where the ρ axis is the real axis.

When the appropriate conditions above are satisfied a system is said to be asymptotically stable: the variables of an asymptotically stable control system always decrease from their initial value and do not show permanent oscillations.

Permanent oscillations occur when a pole has a real part exactly equal to zero (in the continuous time case) or a modulus equal to one (in the discrete time case). If a simply stable system response neither decays nor grows over time, and has no oscillations, it is marginally stable: in this case the system transfer function has non-repeated poles at complex plane origin (*i.e.* their real and complex component is zero in the continuous time case). Oscillations are present when poles with real part equal to zero have an imaginary part not equal to zero.

If a system in question has an impulse response of

$$x[n] = 0.5^n u[n]$$

then the Z-transform, is given by

$$X(z) = \frac{1}{1 - 1.5z^{-1}}$$

which has a pole in $z = 0.5$ (zero imaginary part). This system is BIBO (asymptotically) stable since the pole is *inside* the unit circle.

However, if the impulse response was

$$x[n] = 0.5^n u[n]$$

then the Z-transform is

$$X(z) = \frac{1}{1 - 1.5z^{-1}}$$

which has a pole at $z = 1.5$ and is not BIBO stable since the pole has a modulus strictly greater than one.

Numerous tools exist for the analysis of the poles of a system. These include graphical systems like the root locus, Bode plots or the Nyquist plots.

Mechanical changes can make equipment (and control systems) more stable. Sailors add ballast to improve the stability of ships. Cruise ships use antiroll fins that extend transversely from the side of the ship for perhaps 30 feet (10 m) and are continuously rotated about their axes to develop forces that oppose the roll.

Controllability and Observability

Controllability and observability are main issues in the analysis of a system before deciding the best control strategy to be applied, or whether it is even possible to control or stabilize the system. Controllability is related to the possibility of forcing the system into a particular state by using an appropriate control signal. If a state is not controllable, then no signal will ever be able to control the state. If a state is not controllable, but its dynamics are stable, then the state is termed *stabilizable*. Observability instead is related to the possibility of *observing*, through output measurements, the state of a system. If a state is not observable, the controller will never be able to determine the behaviour of an unobservable state and hence cannot use it to stabilize the system. However, similar to the stabilizability condition above, if a state cannot be observed it might still be detectable.

From a geometrical point of view, looking at the states of each variable of the system to be controlled, every "bad" state of these variables must be controllable and observable to ensure a good behaviour in the closed-loop system. That is, if one of the eigenvalues of the system is not both controllable and observable, this part of the dynamics will remain untouched in the closed-loop system. If such an eigenvalue is not stable, the dynamics of this eigenvalue will be present in the closed-loop system which therefore will be unstable. Unobservable poles are not present in the transfer function realization of a state-space representation, which is why sometimes the latter is preferred in dynamical systems analysis.

Solutions to problems of uncontrollable or unobservable system include adding actuators and sensors.

Control Specification

Several different control strategies have been devised in the past years. These vary from extremely general ones (PID controller), to others devoted to very particular classes of systems (especially robotics or aircraft cruise control).

A control problem can have several specifications. Stability, of course, is always present: the controller must ensure that the closed-loop system is stable, regardless of the open-loop stability. A poor choice of controller can even worsen the stability of the open-loop system, which must normally be avoided. Sometimes it would be desired to obtain particular dynamics in the closed loop: *i.e.* that the poles have $\text{Re}[\lambda] < -\bar{\lambda}$, where $\bar{\lambda}$ is a fixed value strictly greater than zero, instead of simply asking that $\text{Re}[\lambda] < 0$.

Another typical specification is the rejection of a step disturbance; including an integrator in the open-loop chain (*i.e.* directly before the system under control) easily achieves this. Other classes of disturbances need different types of subsystems to be included.

Other "classical" control theory specifications regard the time-response of the closed-loop system: these include the rise time (the time needed by the control system to reach the desired value after a perturbation), peak overshoot (the highest value reached by the response before reaching the desired value) and others (settling time, quarter-decay). Frequency domain specifications are usually related to robustness.

Modern performance assessments use some variation of integrated tracking error (IAE,ISA,CQI).

Model Identification and Robustness

A control system must always have some robustness property. A robust controller is such that its properties do not change much if applied to a system slightly different from the mathematical one used for its synthesis. This specification is important: no real physical system truly behaves like the series of differential equations used to represent it mathematically. Typically a simpler mathematical model is chosen in order to simplify calculations, otherwise the true system dynamics can be so complicated that a complete model is impossible.

System Identification

The process of determining the equations that govern the model's dynamics is called system identification. This can be done off-line: for example, executing a series of measures from which to calculate an approximated mathematical model, typically its transfer function or matrix. Such identification from the output, however, cannot take account of unobservable dynamics. Sometimes the model is built directly starting from known physical equations: for example, in the case of a mass-spring-damper system we know that

$$m\ddot{x}(t) = -Kx(t) - B\dot{x}(t).$$

Even assuming that a "complete" model is used in designing the controller, all the parameters included in these equations (called "nominal parameters") are never known with absolute precision; the control system will have to behave correctly even when connected to physical system with true parameter values away from nominal.

Some advanced control techniques include an "on-line" identification process. The parameters of the model are calculated ("identified") while the controller itself is running: in this way, if a drastic variation of the parameters ensues (for example, if the robot's arm releases a weight), the controller will adjust itself consequently in order to ensure the correct performance.

Analysis

Analysis of the robustness of a SISO (single input single output) control system can be performed in the frequency domain, considering the system's transfer function and using Nyquist and Bode diagrams. Topics include gain and phase margin and amplitude margin. For MIMO (multi input multi output) and, in general, more complicated control systems one must consider the theoretical results devised for each control technique: i.e., if particular robustness qualities are needed, the engineer must shift his attention to a control technique by including them in its properties.

Constraints

A particular robustness issue is the requirement for a control system to perform properly in the presence of input and state constraints. In the physical world every signal is limited. It could happen that a controller will send control signals that cannot be followed by the physical system: for example, trying to rotate a valve at excessive speed. This can produce undesired behavior of the closed-loop system, or even damage or break actuators or other subsystems. Specific control techniques are available to solve the problem: model predictive control, and anti-wind up systems. The latter consists of an additional control block that ensures that the control signal never exceeds a given threshold.

System Classifications

Linear Systems Control

For MIMO systems, pole placement can be performed mathematically using a state space representation of the open-loop system and calculating a feedback matrix assigning poles in the desired positions. In complicated systems this can require computer-assisted calculation capabilities, and cannot always ensure robustness. Furthermore, all system states are not in general measured and so observers must be included and incorporated in pole placement design.

Nonlinear Systems Control

Processes in industries like robotics and the aerospace industry typically have strong nonlinear dynamics. In control theory it is sometimes possible to linearize such classes of systems and apply linear techniques, but in many cases it can be necessary to devise from scratch theories permitting control of nonlinear systems. These, *e.g.*, feedback linearization, backstepping, sliding mode control, trajectory linearization control normally take advantage of results based on Lyapunov's theory. Differential geometry has been widely used as a tool for generalizing well-known linear control concepts to the non-linear case, as well as showing the subtleties that make it a more challenging problem. Control theory has also been used to decipher the neural mechanism that directs cognitive states.

Decentralized Systems Control

When the system is controlled by multiple controllers, the problem is one of decentralized control. Decentralization is helpful in many ways, for instance, it helps control systems to operate over a larger geographical area. The agents in decentralized control systems can interact using communication channels and coordinate their actions.

Deterministic and Stochastic Systems Control

A stochastic control problem is one in which the evolution of the state variables is subjected to random shocks from outside the system. A deterministic control problem is not subject to external random shocks.

Main Control Strategies

Every control system must guarantee first the stability of the closed-loop behavior. For linear systems, this can be obtained by directly placing the poles.

Non-linear control systems use specific theories (normally based on Aleksandr Lyapunov's Theory) to ensure stability without regard to the inner dynamics of the system. The possibility to fulfill different specifications varies from the model considered and the control strategy chosen.

List of the main control techniques:

- Adaptive control uses on-line identification of the process parameters, or modification of controller gains, thereby obtaining strong robustness properties. Adaptive controls were applied for the first time in the aerospace industry in the 1950s, and have found particular success in that field.

- A hierarchical control system is a type of control system in which a set of devices and governing software is arranged in a hierarchical tree. When the links in the tree are implemented by a computer network, then that hierarchical control system is also a form of networked control system.

- Intelligent control uses various AI computing approaches like neural networks, Bayesian probability, fuzzy logic, machine learning, evolutionary computation and genetic algorithms to control a dynamic system.

- Optimal control is a particular control technique in which the control signal optimizes a certain "cost index": for example, in the case of a satellite, the jet thrusts needed to bring it to desired trajectory that consume the least amount of fuel. Two optimal control design methods have been widely used in industrial applications, as it has been shown they can guarantee closed-loop stability. These are Model Predictive Control (MPC) and linear-quadratic-Gaussian control (LQG). The first can more explicitly take into account constraints on the signals in the system, which is an important feature in many industrial processes. However, the "optimal control" structure in MPC is only a means to achieve such a result, as it does not optimize a true performance index of the closed-loop control system. Together with PID controllers, MPC systems are the most widely used control technique in process control.

- Robust control deals explicitly with uncertainty in its approach to controller design. Controllers designed using robust control methods tend to be able to cope with small differences between the true system and the nominal model used for design. The early methods of Bode and others were fairly robust; the state-space methods invented in the 1960s and 1970s were sometimes found to lack robustness. Examples of modern robust control techniques include H-infinity loop-shaping developed by Duncan McFarlane and Keith Glover of Cambridge University, United Kingdom and Sliding mode control (SMC) developed by Vadim Utkin. Robust methods aim to achieve robust performance and/or stability in the presence of small modeling errors.

- Stochastic control deals with control design with uncertainty in the model. In typical stochastic control problems, it is assumed that there exist random noise and disturbances in the model and the controller, and the control design must take into account these random deviations.

- Energy-shaping control view the plant and the controller as energy-transformation devices. The control strategy is formulated in terms of interconnection (in a power-preserving manner) in order to achieve a desired behavior.

- Self-organized criticality control may be defined as attempts to interfere in the processes by which the self-organized system dissipates energy.

TRANSIENT RESPONSE

In electrical engineering and mechanical engineering, a transient response or natural response is the response of a system to a change from equilibrium. The transient response is not necessarily tied to "on/off" events but to any event that affects the equilibrium of the system. The impulse response and step response are transient responses to a specific input (an impulse and a step, respectively).

Damping

The response can be classified as one of three types of damping that describes the output in relation to the steady-state response.

An underdamped response is one that oscillates within a decaying envelope. The more underdamped the system, the more oscillations and longer it takes to reach steady-state. Here damping ratio is always <1.

A critically damped response is the response that reaches the steady-state value the fastest without being underdamped. It is related to critical points in the sense that it straddles the boundary of underdamped and overdamped responses. Here, damping ratio is always equal to one. There should be no oscillation about the steady state value in the ideal case.

An overdamped response is the response that does not oscillate about the steady-state value but takes longer to reach than the critically damped case. Here damping ratio is >1 it is the response of a system with respect to the input as a function of time

Properties

Rise Time

Rise time refers to the time required for a signal to change from a specified low value to a specified high value. Typically, these values are 10% and 90% of the step height.

Overshoot

Overshoot is when a signal or function exceeds its target. It is often associated with ringing.

Settling Time

Settling time is the time elapsed from the application of an ideal instantaneous step input to the time at which the output has entered and remained within a specified error band.

Delay-time

The delay time is the time required for the response to reach half the final value the very first time.

Peak Time

The peak time is the time required for the response to reach the first peak of the overshoot.

Steady-state Error

2003's *Instrument Engineers' Handbook* defines the steady-state error of a system as "the difference between the desired final output and the actual one" when the system reaches a steady state, when its behavior may be expected to continue if the system is undisturbed.

Transient Response, Steady State, and Decay

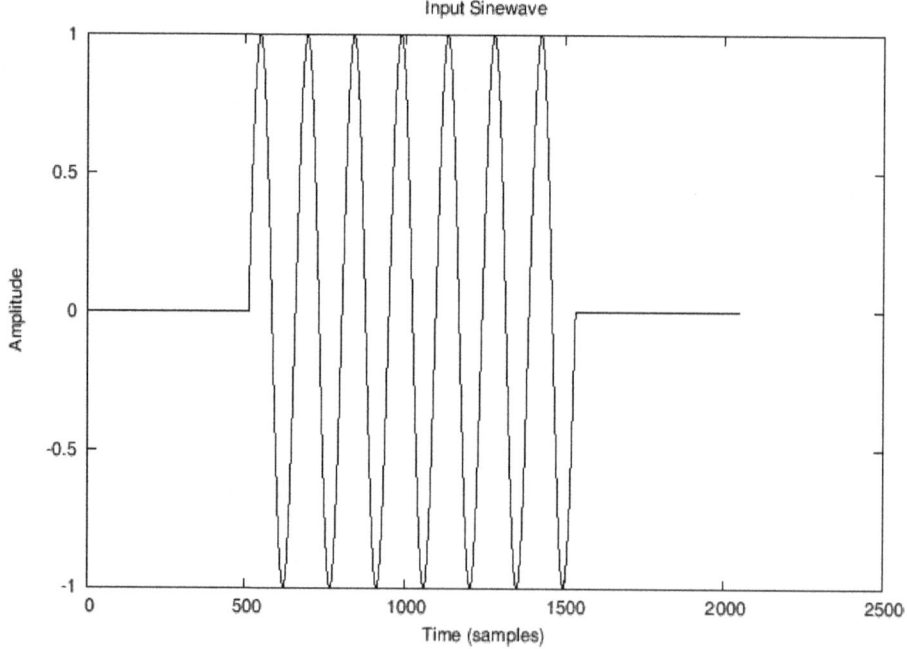

Input Signal

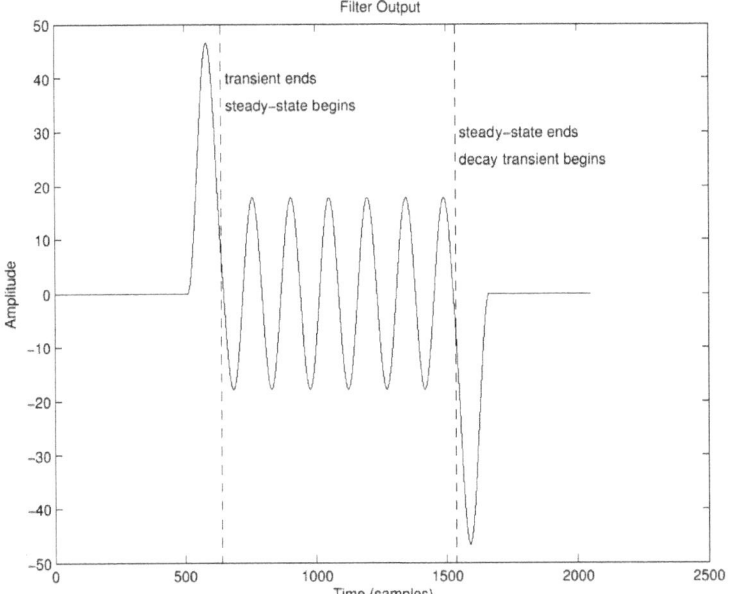

Filter Output Signal

Fig. : Example transient, steady-state, and decay responses for an
FIR "running sum" filter driven by a gated sinusoid.

The terms *transient response* and *steady state response* arise naturally in the context of sinewave analysis. When the input sinewave is switched on, the filter takes a while to "settle down" to a perfect sinewave at the same frequency. The filter response during this "settling" period is called the *transient response* of the filter. The response of the filter *after* the transient response, provided the filter is linear and time-invariant, is called the *steady-state response*, and it consists of a pure sinewave at the same frequency as the input sinewave, but with amplitude and phase determined by the filter's *frequency response* at that frequency. In other words, the steady-state response begins when the LTI filter is fully "warmed up" by the input signal. More precisely, the filter output is the same as if the input signal had been applied since time minus infinity. Length N FIR filters only "remember" $N - 1$ samples into the past. Thus, for length N FIR filters, the duration of the transient response is $N - 1$ samples.

To show this, consider that a length $N = 1$ (zero-order) FIR filter (a simple gain), has no state memory at all, and thus it is in "steady state" immediately when the input sinewave is switched on. A length $N = 2$ FIR filter, on the other hand, reaches steady state one sample after the input sinewave is switched on, because it has one sample of delay. At the switch-on time instant, the length 2 FIR filter has a single sample of state that is still zero (instead of its steady-state value which is the previous input sinewave sample).

In general, a length N FIR filter is fully "warmed up" after $N - 1$ samples of input; that is, for an input starting at time $n = 0$, by time n $= N - 1$, all internal state delays of the filter contain delayed input samples instead of their initial zeros. When the input signal is a unit step $u(n)$ times a sinusoid (or, by superposition, any linear combination of sinusoids), we may say that the filter output reaches *steady state* at time $n = N - 1$.

PROPORTIONAL CONTROL

A proportional control system is a type of linear feedback control system. Two classic mechanical examples are the toilet bowl float proportioning valve and the fly-ball governor.

The proportional control system is more complex than an on-off control system like a bi-metallic domestic thermostat, but simpler than a proportional-integral-derivative (PID) control system used in something like an automobile cruise control. On-off control will work where the overall system has a relatively long response time, but can result in instability if the system being controlled has a rapid response time. Proportional control overcomes this by modulating the output to the controlling device, such as a continuously variable valve.

An analogy to on-off control is driving a car by applying either full power or no power and varying the duty cycle, to control speed. The power would be on until the target speed is reached, and then the power would be removed, so the car reduces speed. When the speed falls below the target, with a certain hysteresis, full power would again be applied. It can be seen that this looks like pulse-width modulation, but would obviously result in poor control and large variations in speed. The more powerful the engine; the greater the instability, the heavier the car; the greater the stability. Stability may be expressed as correlating to the power-to-weight ratio of the vehicle.

Proportional control is how most drivers control the speed of a car. If the car is at target speed and the speed increases slightly, the power is reduced slightly, or in proportion to the error (the actual versus target speed), so that the car reduces speed gradually and reaches the target point with very little, if any, "overshoot", so the result is much smoother control than on-off control.

Further refinements like PID control would help compensate for additional variables like hills, where the amount of power needed for a given speed change would vary. This would be accounted for by the integral function of the PID control.

Proportional Control Theory

In the proportional control algorithm, the controller output is proportional to the error signal, which is the difference between the set point and the process variable. In other words, the output of a proportional controller is the multiplication product of the error signal and the proportional gain.

This can be mathematically expressed as

$$P_{out} = K_p\, e(t) + p0$$

where

- $p0$: Controller output with zero error.
- P_{out} : Output of the proportional controller
- K_p : Proportional gain
- $e(t)$: Instantaneous process error at time t. $e(t) = SP - PV$
- SP: Set point
- PV: Process variable

Offset Error

Proportional Control Action leaves out an error called Offset Error. Once a disturbance (deviation from existing state) occurs in the Steady State Condition, any corrective control action, based purely on Proportional Control, will always leave out an error between the next steady state and the desired set point. This error is called an Offset Error.

PROPORTIONAL INTEGRAL AND DERIVATIVE CONTROLLERS

Under Control System

Before I introduce you about various controllers, it is very essential to know the uses of controllers in the theory of control systems. The important uses of the controllers are written below:

1. Controllers improve steady state accuracy by decreasing the steady state errors.
2. As the steady state accuracy improves, the stability also improves.
3. They also help in reducing the offsets produced in the system.
4. Maximum overshoot of the system can be controlled using these controllers.
5. They also help in reducing the noise signals produced in the system.
6. Slow response of the over damped system can be made faster with the help of these controllers.

Now what are controllers? A controller is one which compares controlled values with the desired values and has a function to correct the deviation produced.

Types of Controllers

Let us classify the controllers. There are mainly two **types of controllers** and they are written below: **Continuous Controllers:** The main feature of continuous controllers is that the controlled variable (also known as the manipulated variable) can have any value within the range of controller's output. Now in the continuous controller's theory, there are three basic modes on which the whole control action takes place and these modes are written below. We will use the combination of these modes in order to have a desired and accurate output.

1. **Proportional controllers.**
2. **Integral controllers.**
3. **Derivative controllers.**

Combinations of these three controllers are written below:

4. Proportional and integral controllers.
5. Proportional and derivative controllers.

Proportional Controllers

We cannot use **types of controllers** at anywhere, with each type controller, there are certain conditions that must be fulfilled. With **proportional controllers** there are two conditions and these are written below:

1. Deviation should not be large, it means there should be less deviation between the input and output.
2. Deviation should not be sudden.

Now we are in a condition to discuss proportional controllers, as the name suggests in a proportional controller the output (also called the actuating signal) is directly proportional to the error signal. Now let us analyze proportional controller mathematically. As we know in proportional controller output is directly proportional to error signal, writing this mathematically we have,

$$A(t) \propto e(t)$$

Removing the sign of proportionality we have,

$$A(t) = K_p \times e(t)$$

Where K_p is proportional constant also known as controller gain. It is recommended that K_p should be kept greater than unity. If the value of K_p is greater than unity, then it will amplify the error signal and thus the amplified error signal can be detected easily.

Advantages of Proportional Controller

Now let us discuss some advantages of proportional controller.

1. Proportional controller helps in reducing the steady state error, thus makes the system more stable.
2. Slow response of the over damped system can be made faster with the help of these controllers.

Disadvantages of Proportional Controller

Now there are some serious disadvantages of these controllers and these are written as follows:

1. Due to presence of these controllers we some offsets in the system.
2. Proportional controllers also increase the maximum overshoot of the system.

Integral Controllers

As the name suggests in **integral controllers** the output (also called the actuating signal) is directly proportional to the integral of the error signal. Now let us analyze integral controller mathematically. As we know in an integral controller output is directly proportional to the integration of the error signal, writing this mathematically we have,

$$A(t) \propto \int_0^t e(t)\,dt$$

Removing the sign of proportionality we have,

$$A(t) = K_i \times \int_0^t e(t)\,dt$$

Where K_i is integral constant also known as controller gain. Integral controller is also known as reset controller.

Advantages of Integral Controller

Due to their unique ability they can return the controlled variable back to the exact set point following a disturbance that's why these are known as reset controllers.

Disadvantages of Integral Controller

It tends to make the system unstable because it responds slowly towards the produced error.

Derivative Controllers

We never use **derivative controllers** alone. It should be used in combinations with other modes of controllers because of its few disadvantages which are written below:

1. It never improves the steady state error.
2. It produces saturation effects and also amplifies the noise signals produced in the system.

Now, as the name suggests in a derivative controller the output (also called the actuating signal) is directly proportional to the derivative of the error signal. Now let us analyze derivative controller mathematically. As we know in a derivative controller output is directly proportional to the derivative of the error signal, writing this mathematically we have,

$$A(t) \propto \frac{de(t)}{dt}$$

Removing the sign of proportionality we have,

$$A(t) = K_d \times \frac{de(t)}{dt}$$

Where K_d is proportional constant also known as controller gain. Derivative controller is also known as rate controller.

Advantages of Derivative Controller

The major advantage of derivative controller is that it improves the transient response of the system.

Proportional and Integral Controller

As the name suggests it is a combination of proportional and an integral controller the output (also called the actuating signal) is equal to the summation of proportional and integral of the error signal. Now let us analyze proportional and integral controller mathematically. As we know in a proportional and integral controller output is directly proportional to the summation of proportional of error and integration of the error signal, writing this mathematically we have,

$$A(t) \propto \int_0^t e(t)\, dt + A(t) \propto e(t)$$

Removing the sign of proportionality we have,

$$A(t) = K_i \int_0^t e(t)\, dt + K_p e(t)$$

Where K_i and k_p proportional constant and integral constant respectively.

Advantages and disadvantages are the combinations of the advantages and disadvantages of proportional and integral controllers.

Proportional and Derivative Controller

As the name suggests it is a combination of proportional and a derivative controller the output (also called the actuating signal) is equals to the summation of proportional and derivative of the error signal. Now let us analyze proportional and derivative controller mathematically. As we know in a proportional and derivative controller output is directly proportional to summation of proportional of error and differentiation of the error signal, writing this mathematically we have,

$$A(t) \propto \frac{de(t)}{dt} + A(t) \propto e(t)$$

Removing the sign of proportionality we have,

$$A(t) = K_d \frac{de(t)}{dt} + K_p e(t)$$

Where K_d and k_p proportional constant and derivative constant respectively.

Advantages and disadvantages are the combinations of advantages and disadvantages of proportional and derivative controllers

BANG-BANG *VS.* PROPORTIONAL CONTROL

Our thermostat, regardless of whether it's wired for negative or positive feedback, applies an all or nothing form of control. When the temperature drops too low, the furnace comes on, Hell bent, to heat up the place. When it gets too hot, the air conditioner's unleashed, flat out, to chill out the environs.

Engineers call this kind of control *"bang-bang."* When it gets too cold *bang*, the furnace comes on. As soon as the temperature rises above 60°, *bang* the furnace cuts off. Too hot? *Bang*, the air conditioner starts, and so on.

Take a closer look at the inside temperature. The temperature tends to oscillate between 70 degrees and the level where the thermostat kicks in? That's the signature of bang-bang control--since nothing happens until one of the limits is hit, the temperature varies freely between them. When the system exceeds a limit, it's hauled back within range, then allowed to drift again.

Bang-bang control keeps the temperature pretty much within the range from 60 to 80, but it allows the temperature to vary freely between the limits. Suppose instead of just switching the furnace and air conditioner on and off, we coupled the temperature reading to the gas valve on the furnace: the further the temperature falls below 70, the more heat the furnace generates. Likewise, as soon as the temperature rises above 70, the air conditioner starts, but we rig it to generate more and more cooling as the temperature rises. We'll end up with a system that behaves like this.

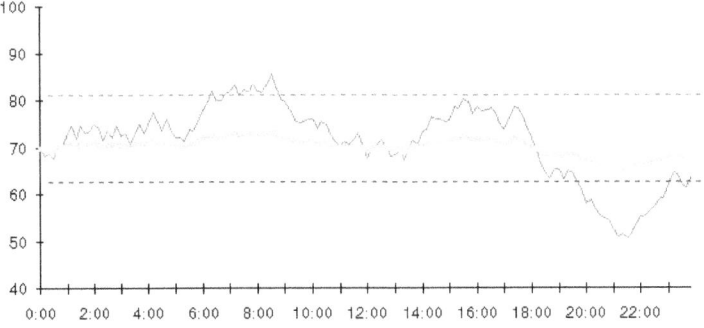

This is called *proportional control*. The action taken, the feedback, is *in proportion* to the degree the system diverges from the ideal point. Imagine you're driving down the road some lonely night. Proportional control is how you usually drive; every time you notice the car drifting a tiny bit to the left, you steer slightly to the right and vice versa. Bang-bang control would mean ignoring the steering wheel until the car crossed a lane marker line. Then you'd haul it in the opposite direction until the car was no longer outside the lane: exciting, perhaps, but not recommended.

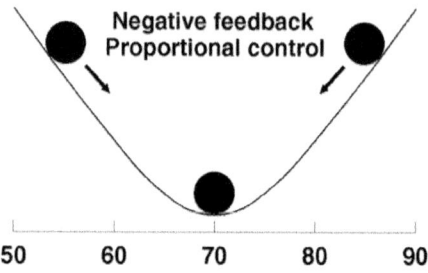

If bang-bang negative feedback keeps the temperature ball on the flat floor of a valley with steep sides, proportional control confines it to the bottom of a smoothly sloped bowl. The slightest degree of motion away from the optimal point, the bottom of the bowl, causes the ball to roll upward. The further it deviates from the goal, the stronger the force applied to restore it--back to the bottom of the bowl. A chart of positive feedback with proportional control would simply flip the bowl upside down: the temperature would remain stable only as long as it stayed perched at the precise optimum point. To appreciate how quickly proportional positive feedback gets out of hand, try balancing a raw egg on top of a bowling ball.

Proportional control makes systems run smoother than bang-bang. Most biological systems are proportional, while many engineered and all too many political and social systems are bang-bang. As with negative and positive feedback, once you understand how proportional control lends stability to a system, while bang-bang tends to oscillate between the extremes, you'll recognise many examples of both kinds of control in everyday life.

Chapter 2

CONTROL OF INTEGRATING PROCESSES

RECOGNIZING INTEGRATING (NON-SELF REGULATING) PROCESS BEHAVIOR

The case studies on this site largely focus on the control of self regulating processes. The principal characteristic that makes a process self regulating is that it naturally seeks a steady state operating level if the controller output and disturbance variables are held constant for a sufficient period of time.

Cruise control of a car is a self regulating process. If we keep the fuel flow to the engine constant while traveling on flat ground on a windless day, the car will settle out at some constant speed. If we increase the fuel flow rate a fixed amount, the car will accelerate and then steady out at a different constant speed.

The heat exchanger process that has been studied on this site is self regulating. If the exchanger cooling rate and disturbance flow rate are held constant at fixed values, the exit temperature will steady at a constant value. If we increase the cooling rate, wait a bit, and then return it to its original value, the exchanger exit temperature will respond during the experiment and then return to its original steady state.

But some processes where the streams are comprised of gases, liquids, powders, slurries and melts do not naturally settle out at a steady state operating level. Process control practitioners refer to these as non-self regulating, or more commonly, as integrating processes.

Integrating (non-self regulating) processes can be remarkably challenging to control. After exploring the distinctive behaviors illustrated below, you may come to realize that some of the level, temperature, pressure, pH and other loops you work with have such a character.

Integrating (Non-Self Regulating) Behavior in Manual Mode

The upper plot below shows the open loop (manual mode) behavior of the more common self regulating process. In this idealized response, the controller output (CO) and process variable (PV) are initially at steady state. The CO is stepped up and back from this steady state. As shown, the PV responds to the step, but ultimately returns to its original operating level.

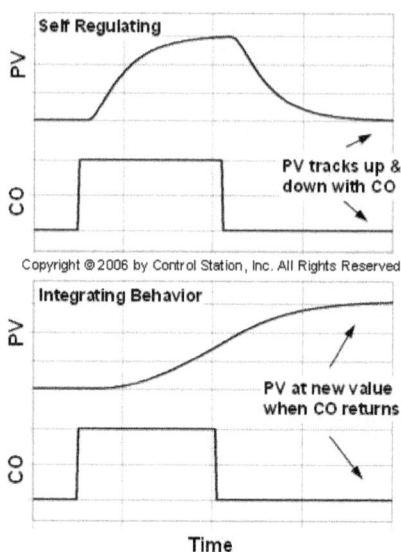

Copyright © 2006 by Control Station, Inc. All Rights Reserved.

The lower plot above shows the open loop response of an ideal integrating process. The distinctive behavior is that the PV settles at a new operating level when the CO returns to its original value.

In truth, the integrating behavior plot above is misleading in that it implies that for such processes, a steady CO will produce a steady PV. While possible with idealized simulations like that used to generate the plot, such a "balance point" behavior is rarely found in open loop (manual mode) for integrating processes in an industrial operation.

More realistically, if left uncontrolled, the lack of a balance point means the PV of an integrating process will naturally tend to drift up or down, possibly to extreme and even dangerous levels. Consequently, integrating processes are rarely operated in manual mode for very long.

Aside: the behavior shown in the integrating plot can also appear across small regions of operation of a self-regulating process if there is a significant dead-band in the final control element (FCE). This might result, for example, from loose mechanical linkages in a valve. For this investigation, we assume the FCE operates properly.

P-Only Control Behavior is Different

To appreciate the difference in behavior for integrating processes, we first recall P-Only control of a self regulating process.

The set point (SP) is initially at the design level of operation (DLO) in the first moments of operation, then PV equals SP. Recall that the DLO is where we expect the SP and PV to be during normal operation when the major disturbances are at their normal or typical values.

The set point is then stepped up from the DLO on the left half of the plot. The simple P-Only controller is unable to track the changing SP and a steady error, called offset, results. The offset grows as each step moves the SP farther away from the DLO.

Midway through the plot, a disturbance occurs. Its size was pre-calculated in this ideal simulation to eliminate the offset. The SP is then stepped back down to the right in the plot and we see that offset shifts, but again grows in a similar and predictable pattern.

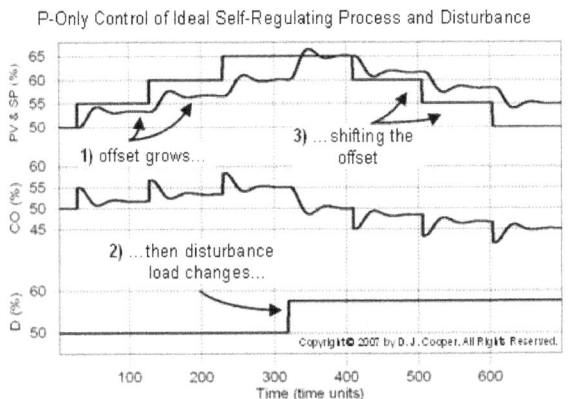

With this as background, we next consider an ideal integrating process simulation under P-Only control.

Even under simple P-Only control as shown in the left half of the plot, the PV is able to track the SP steps with no offset. This behavior can be quite confusing as it does not fit the expected behavior we have just seen above for the more common self regulating process.

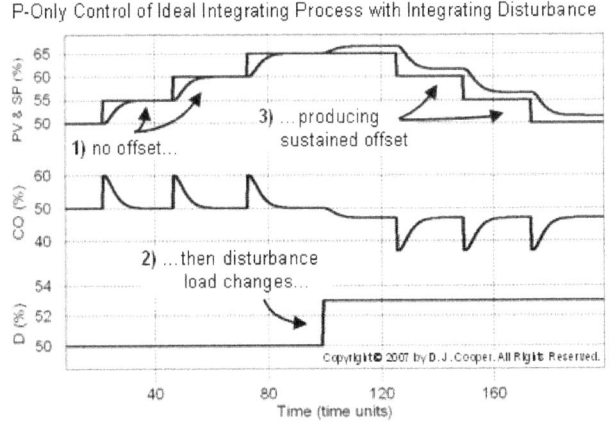

The reason this happens is that integrating processes have a natural accumulating character. In fact, this is why "integrating process" is used as a descriptor for non-self regulating processes. Since the process integrates, then it appears that the controller does not need to.

Yet the set point steps to the right in the above plot show that this is not completely correct. Once a disturbance shifts the baseline operation of the process, shown roughly at the midpoint in the above plot, an offset develops and remains constant, even as SP returns to its original design value.

Controller Output Behavior is Telling

If we study the CO trace in the two plots above, we see one feature that distinguishes self regulating from integrating process behavior.

In the self regulating process plot above, the average CO value tracks up and then down as the SP steps up and then down.

In the integrating process plot above, the CO spikes with each SP step, but then in a most unintuitive fashion, returns to the same steady value. It is only the change in the disturbance flow that causes the average CO to shift midway through the plot, though it then remains centered around the new value for the remainder of the SP steps.

PI Control Behavior is Different

The ideal self regulating process controlled using the popular dependent ideal PI algorithm. Reset time, T_i, is held constant throughout the experiment while controller gain, K_c, is doubled and then doubled again.

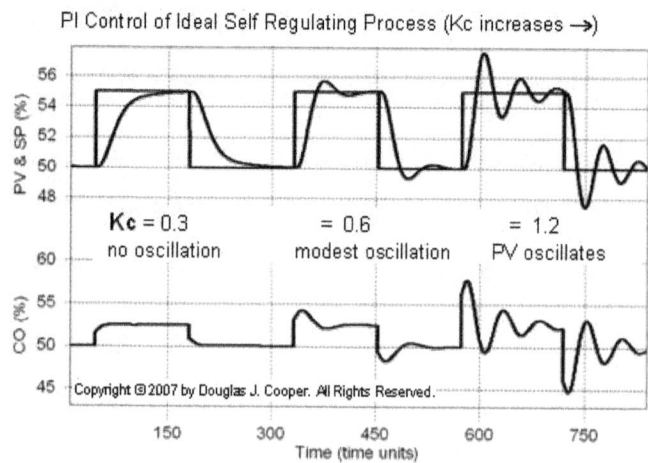

As Kc increases, the controller becomes more active, and as we have grown to expect, this increases the tendency of the PV to display oscillating (or underdamped) behavior.

For comparison, now consider PI control of an ideal integrating process simulation. As above, reset time, T_i, is held constant throughout the experiment while controller gain, K_c, is increased across the plot.

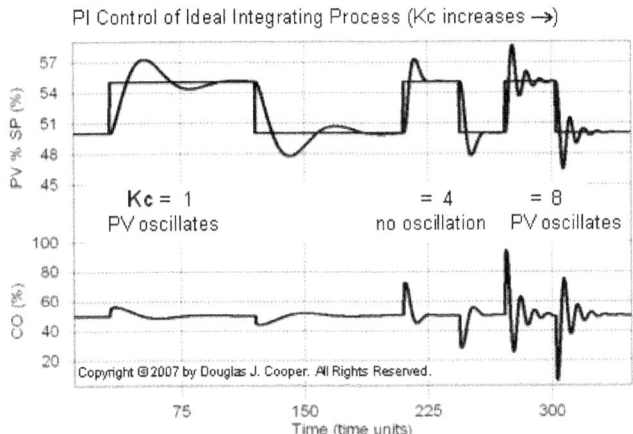

A counter-intuitive result is that as K_c becomes small *and* as it becomes large, the PV begins displaying an underdamped (oscillating) response behavior.

While the frequency of the oscillations is clearly different between a small and large K_c when seen together in a single plot as above, it is not always obvious what direction controller gain needs to move to settle the process when looking at such unacceptable performance on a control room display.

Tuning Recipe Required

One of the biggest challenges for practitioners is recognizing that a particular process shows integrating behavior prior to starting a controller design and tuning project. This, like most things, comes with training, experience and practice.

Once in automatic, closed loop behavior of an integrating process can be unintuitive and even confounding. Trial and error tuning can lead us in circles as we try to understand what is causing the problem.

A formal controller design and tuning recipe for integrating processes helps us overcome these issues in an orderly and reliable fashion.

A DESIGN AND TUNING RECIPE FOR INTEGRATING PROCESSES

This e-book, it is best practice to follow a formal recipe when designing and tuning a PID controller. A recipe lets us move a controller into operation quickly. And perhaps most important, the performance of the controller will be superior to one tuned using an intuitive approach or trial-and-error method.

Additionally, a recipe-based approach overcomes many of the concerns that makes control projects challenging in an industrial operating environment. Specifically, a recipe approach causes less disruption to the production schedule, wastes less raw material and utilities, requires less personnel time, and generates less off-spec product.

The Recipe for Integrating Processes

Integrating (or non-self regulating) processes display counter-intuitive behaviors that make them surprisingly challenging to control. In particular, they do not naturally settle out to a steady operating level if left uncontrolled.

So while the controller design and tuning recipe is generally the same for both self regulating and integrating processes, there are important differences. Specifically, step 3 of the recipe uses a different dynamic model form and step 4 employs different tuning correlations.

Yet the design and tuning recipe maintains the familiar four step structure:

1. Establish the design level of operation (the normal or expected values for set point and major disturbances).
2. Bump the process and collect controller output (CO) to process variable (PV) dynamic process data around this design level.
3. Approximate the process data behavior with a *first order plus dead time integrating* (FOPDT Integrating) dynamic model.
4. Use the model parameters from step 3 in rules and correlations to complete the controller design and tuning.

It is important to recognize that real processes are more complex than the simple FOPDT Integrating model form used in step 3. In spite of this, the FOPDT Integrating model succeeds in providing an approximation of process behavior that is sufficiently accurate to yield reliable and predictable control performance when used with the rules and correlations in step 4 of the recipe.

The FOPDT Integrating Model

We recall that the familiar first order plus dead time (FOPDT) dynamic model used to approximate self regulating dynamic process behavior has the form:

$$Tp\frac{dPV(t)}{dt} + PV(t) = Kp \cdot CO(t - \theta p) \quad \textit{FOPDT Form}$$

Yet this model cannot describe the kind of integrating process behavior shown in these examples. Such behavior is better described with the **FOPDT Integrating model** form:

$$\frac{dPV(t)}{dt} = Kp^* \cdot CO(t - \theta p) \quad \textit{FOPDT Integrating Form}$$

It is interesting to note when comparing the two models above that the FOPDT Integrating form does not have the lone "+ PV" term found on the left hand side of the FOPDT dynamic model.

Also, individual values for the familiar process gain, K_p, and process time constant, T_p, are not separately identified for the FOPDT Integrating model. Instead, an integrator gain, K_p^*, is defined that has units of the ratio of the process gain to the process time constant, or:

$$Kp^*[=]\frac{Kp}{Tp} \quad \text{or} \quad K_p^* [=] PV/(CO.time)$$

Tuning Correlations for Integrating Processes

Analogous to the FOPDT investigations on this site, we will see that the FOPDT Integrating model parameters Kp* and θp of Step 3 can be computed using a graphical analysis of plot data or by automated analysis using commercial software.

Step 4 then provides tuning values for controllers such as the dependent, ideal PI form:

$$CO = CO_{bias} + Kc \cdot e(t) + \frac{Kc}{Ti} \int e(t)\, dt$$

and the dependent, ideal PID form:

$$CO = CO_{bias} + Kc \cdot e(t) + \frac{Kc}{Ti} \int e(t)\, dt - Kc \cdot Td \frac{dPV}{dt}$$

One important difference about integrating processes is that since there is no identifiable process time constant in the FOPDT Integrating model, we use dead time, θp, as the baseline marker of time in the design and tuning rules.

Specifically, θ is used as the basis for computing sample time, T, and the closed loop time constant, Tc. Following the procedures widely discussed on this site for self regulating processes, we employ a rule to compute the closed time constant, Tc, as:

$$Tc = 3\,\theta p$$

The controller tuning correlations for integrating processes use this Tc, as well as the Kp* and θp from the FOPDT integrating model fit, as:

	Controller Gain Kc	Reset Time Ti	Deriv Time Td
PI	$\dfrac{1}{Kp^*}\dfrac{2Tc+\theta p}{(Tc+\theta p)^2}$	$2Tc+\theta p$	
PID	$\dfrac{1}{Kp^*}\dfrac{2Tc+\theta p}{(Tc+0.5\theta p)^2}$	$2Tc+\theta p$	$\dfrac{0.25\theta p^2 + Tc\theta p}{2Tc+\theta p}$

Loop Sample Time, T

Determining a proper sample time, T, for integrating processes is somewhat more challenging than for self regulating processes.

There are two sample times, T, used in process controller design and tuning. One is the control loop sample time that specifies how often the controller samples the measured process variable (PV) and computes and transmits a new controller output (CO) signal. The other is the rate at which CO and PV data are sampled and recorded during a bump test (step 2 of the recipe).

All controllers measure, act, then wait until next sample time before repeating the loop. This "measure, act, wait" procedure has a delay (or dead time) of one sample time built naturally into its structure. Thus, the minimum dead time (θp, min) in any control loop is the loop sample time, T.

Using the Recipe

The tuning recipe for integrating processes has important differences from that used for self regulating process. When designing and tuning controllers for such processes, we should:

- use an FOPDT Integrating model form when approximating dynamic model behavior,
- note that the closed loop time constant, Tc, and sample time, T, are based on model dead time, θp.
- employ PI and PID tuning correlations specific to integrating processes.

ANALYZING PUMPED TANK DYNAMICS WITH A FOPDT INTEGRATING MODEL

Integrating (or non-self regulating) processes display counter-intuitive behaviors that make them surprisingly challenging to control. In particular, they do not naturally settle out to a steady operating level if left uncontrolled.

To address this distinctive dynamic character, we modify the controller design and tuning recipe to include a FOPDT Integrating model and slightly different design rules and tuning correlations.

The Pumped Tank Process

To better understand the design and tuning of a PID controller for an integrating process, we explore the pumped tank case study from Control Station's Loop-Prosoftware.

The process has two liquid streams feeding the top of the tank and a single exit stream pumped out the bottom.

The measured process variable (PV) is liquid level in the tank. To maintain level, the controller output (CO) signal adjusts a throttling valve at the discharge of a constant pressure pump to manipulate flow rate out of the bottom of the tank. This approximates the behavior of a centrifugal pump operating at relatively low throughput.

LOOP-PRO's Pumped Tank in Manual Mode

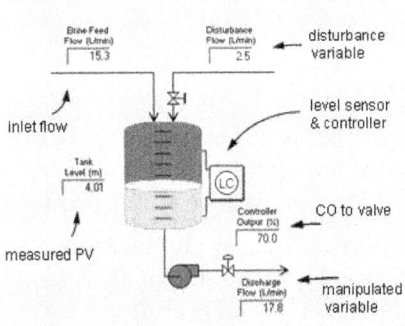

Copyright © 2007 by Douglas J. Cooper. All Rights Reserved.

Unlike the gravity drained tanks case study where the exit flow rate increases and decreases as tank level rises and falls, the discharge flow rate here is strictly regulated by a pump. As a consequence, the physics do not naturally work to balance the system when any of the stream flow rates change.

This lack of a natural balancing behavior is why the pumped tank is classified as an integrating process. If the total flow into the tank is greater than the flow pumped out, the liquid level will rise and continue to rise until the tank fills or a stream flow changes. If the total flow into the tank is less than the flow pumped out, the liquid level will fall and continue to fall.

Below is a plot of the pumped tank behavior with the controller in manual mode (open loop). The CO signal is stepped up, increasing the discharge flow rate out of the bottom of the tank. The flow out becomes greater that the total feed into the top of the tank and as shown, liquid level begins to fall.

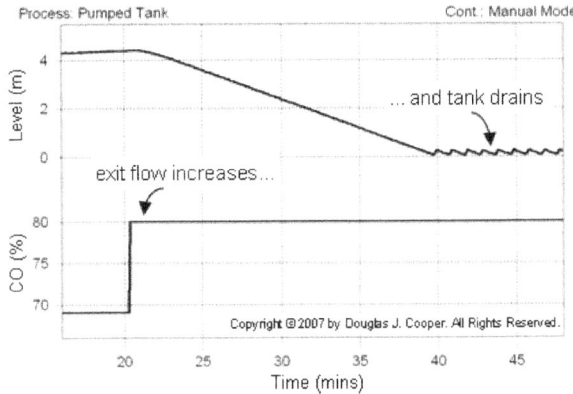

As the situation persists, liquid level continues to fall until the tank is drained. The saw-toothed effect shown when the tank is empty is because the pump briefly surges every time enough liquid accumulates for it to regain suction.

Not shown is that if the controller output were to be decreased enough to cause flow rate out to be less than flow rate in, the tank level would rise until full.

If this were a real process, the tank would overflow and spill, creating safety and profitability issues.

The Disturbance Stream

The disturbance variable is a flow rate of a secondary feed into the top of the tank. This disturbance flow (D) is controlled independently, as if by another process (which is why it is a disturbance to our process).

When D decreases (or increases), the measured PV level falls (or rises) in response. To illustrate, the plot below shows that the CO is held constant and D is decreased. Characteristic of an integrating process, the PV (tank level) starts falling because total flow into the tank is less than that pumped out.

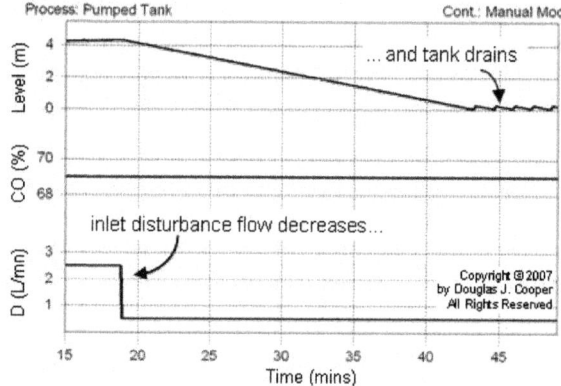

Pumped Tank in Closed Loop

The process graphic below shows the pumped tank in automatic mode (closed loop). The two streams feeding into the top of the tank total 17.8 L/min and level is steady because the controller regulates the discharge flow to this same value.

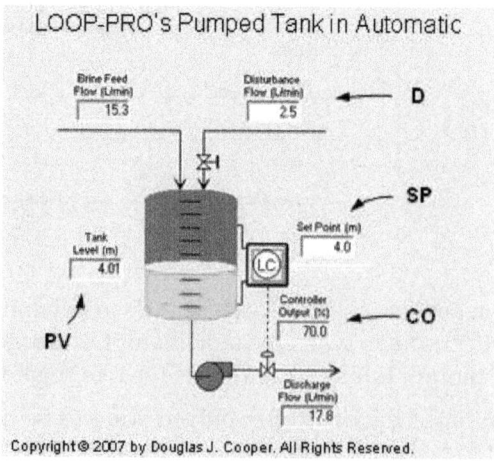

Copyright © 2007 by Douglas J. Cooper. All Rights Reserved.

Graphical Modeling of Integrating Process Data

When collecting and analyzing data as detailed in steps 1 and 2 of the design and tuning recipe for integrating processes, we begin by specifying a design level of operation (DLO). In this study, we specify:

Design PV = 4.8 m

Design D = 2.5 L/min

Note that while not shown in the plots, D is held constant at 2.5 L/min throughout this study.

Just as with self regulating processes, step 3 of the recipe uses a simplifying dynamic model approximation to describe the complex behavior of a real process. The FOPDT Integrating model is simple in form yet it provides information sufficiently accurate for controller design and tuning.

The FOPDT Integrating model is:

$$\frac{dPV(t)}{dt} = Kp^* \cdot CO(t - \theta p)$$

where the integrator gain, Kp^*, has units of the ratio of the process gain to the process time constant, or:

$$Kp^*[=]\frac{Kp}{Tp} \quad \text{or} \quad Kp^* [=] PV/(CO.time)$$

The graphical method of fitting a FOPDT Integrating model to process data requires a data set that includes at least two constant values of controller output, CO_1 and CO_2.

The pumped tank, both must be held constant long enough so that a slope trend in the PV response (tank liquid level) can be visually identified.

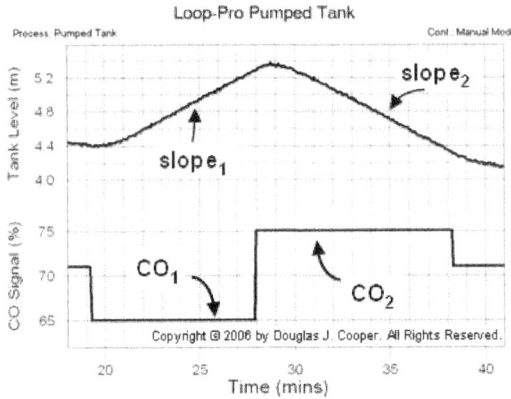

An important difference between the graphical technique for self-regulating processes and integrating processes is that integrating processes need not start at a steady value (steady-state) before a bump is made to the CO. The graphical technique is only concerned with the slopes (or rates of change) in PV and the controller output signal that caused each PV slope.

The FOPDT Integrating model describes the PV behavior at each value of constant controller output CO_1 and CO_2 as:

$$\left.\frac{dPV}{dt}\right|_1 = Kp^* CO_1(t - \theta p)$$

and

$$\left.\frac{dPV}{dt}\right|_2 = Kp^* CO_2(t - \theta p)$$

Subtracting and solving for Kp^* yields:

$$Kp^* = \frac{\left.\frac{dPV}{dt}\right|_2 - \left.\frac{dPV}{dt}\right|_1}{CO_2 - CO_1} = \frac{slope_2 - slope_1}{CO_2 - CO_1}$$

Graphical Modeling of Pumped Tank Data

Computing Integrator Gain

Below is the same open loop data from the pumped tank simulation as shown above. The CO is stepped from 71% down to 65%, causing the liquid level (the PV) to rise. The controller output is then stepped from 65% up to 75%, causing a downward slope in the liquid level.

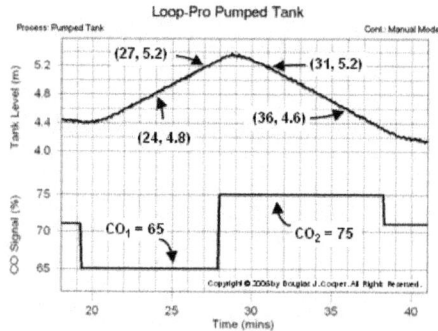

The slope of each segment is calculated as the change in tank liquid level divided by the change in time. From the plot data we compute:

$$slope_1 = \left.\frac{dPV}{dt}\right|_1 = \frac{dPV_1}{\Delta t_1} = \frac{5.2 - 4.8}{27 - 24} = 0.13 \frac{m}{min}$$

and

$$slope_2 = \left.\frac{dPV}{dt}\right|_2 = \frac{dPV_2}{\Delta t_2} = \frac{4.6 - 5.2}{36 - 31} = -0.12 \frac{m}{min}$$

Using the two slopes computed above along with their respective CO values from the plot yields the integrator gain, Kp*, for the pumped tank:

$$Kp^* = \frac{slope_2 - slope_1}{CO_2 - CO_1} = \frac{(-0.12) - (0.13)}{75 - 65} = -0.025 \frac{m}{\% \, min}$$

Computing Dead Time

The dead time is estimated from the plot using the same method described for the heat exchanger. That is, dead time, θp, is computed as the difference in time from when the CO signal was stepped and when the measured PV starts a clear response to that change.

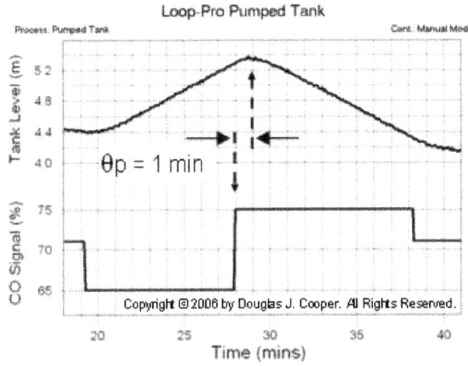

As shown above, the pumped tank dead time is estimated from the plot as:

θp = 1.0 min

Thus, the FOPDT Integrator model information needed to proceed with controller tuning using the correlations presented here is complete.

Automating the Model Fit

In today's world, there is no need to perform the model fit with graph paper and calculator. Commercial software offers analysis tools that makes fitting an FOPDT Integrating model quite simple.

Below, Control Station's Loop-Pro software is used to analyze the pumped tank data. As shown the software displays the data as a plot that includes adjustable nodes and tie lines.

To compute a fit, click on the two CO nodes at the bottom of the plot and drag them to match the two values of constant controller output expected in the data. Both have tie lines to identify their associated PV slope bar. Each slope bar has two end point nodes. Click and drag these so that each bar approximates the sloping segments on the graph.

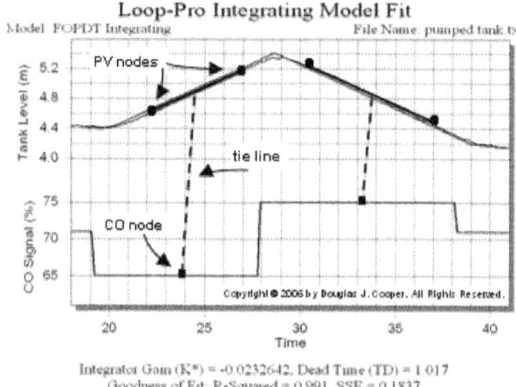

Loop-Pro Integrating Model Fit

Model FOPDT Integrating   File Name: pumped tank.tx

Integrator Gain (K*) = -0.0232642, Dead Time (TD) = 1.017
Goodness of Fit: R-Squared = 0.991, SSE = 0.1837

With the six nodes (two CO and four PV) properly positioned, the FOPDT Integrating model parameters are automatically calculated and the model fit is displayed over the raw data.

We can see that the model PV line matches the measured PV data, so we have confidence that the model fit is good. The image above also shows the FOPDT Integrator values computed by the software:

$$Kp^* = -0.023 \text{ m}/(\% \text{ min}), \quad \text{-p} = 1.0 \text{ min}$$

Recall that earlier in this article we computed by hand:

$$Kp^* = -0.025 \text{ m}/(\% \text{ min}), \quad \text{-p} = 1.0 \text{ min}$$

In the dynamic modeling world, these values are virtually identical. We note that software offers additional benefits in that it performs the computation quickly, reduces the chance of computational error, and provides a visual confirmation that the model used for controller design and tuning reasonably matches the process data.

PI Control Study

With bump test data centered around our design level of operation, and with an approximating FOPDT Integrating model of that data computed, we have completed steps 1-3 of the controller design and tuning recipe for this integrating process.

We complete the study by exploring PI control for the pumped tank process.

PI CONTROL OF THE INTEGRATING PUMPED TANK PROCESS

The control objective for the pumped tank process is to maintain liquid level at set point by adjusting the discharge flow rate out of the bottom of the tank. This process displays the distinctive integrating (or non-self regulating)behavior, and as such, presents an interesting control challenge.

The process graphic below shows the pumped tank in automatic mode (also called closed loop):

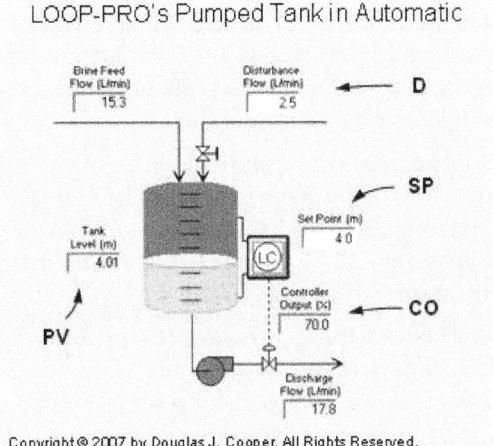

LOOP-PRO's Pumped Tank in Automatic

Copyright © 2007 by Douglas J. Cooper. All Rights Reserved.

The important variables for this study are labeled in the above graphic:

CO = signal to valve that adjusts discharge flow rate of liquid (controller output, %)

PV = measured liquid level signal from the tank (measured process variable, m)

SP = desired liquid level in tank (set point, m)

D = flow rate of liquid entering top of tank (major disturbance, L/min)

We follow the controller design and tuning recipe for integrating processes in this study as we design and test a PI controller. Please recall that there are subtle yet important differences between this procedure and the design and tuning recipe used for the more common self regulating process.

Step 1: Design Level of Operation (DLO)

We choose the DLO to be the same as that used in this article so we can build upon our previous modeling efforts:

design value for PV and SP = 4.8 m

design value for D = 2.5 L/min

Characteristic of real integrating processes, the pumped tank PV does not naturally settle at a steady operating level if CO is held constant. This lack of a natural "balance point" means we will not specify a CO as part of our DLO.

Step 2: Collect Process Data around the DLO

When PV and D are near the design level of operation and D is substantially quiet, we perform a dynamic test and generate CO-to-PV cause and effect dynamic data.

Because the PV of integrating processes tends to drift in manual mode (open loop), one alternative is to perform an open loop dynamic test that does not require bringing the process to steady state. The procedure is to maintain CO at a constant value until a slope trend in the PV can be visually identified. We then move the CO to a different value and hold it until a second PV slope is established. To analyze the resulting plot data with hand calculations to obtain the FOPDT Integrating model for step 3 of the design and tuning recipe.

An alternative approach, presented below, is to use automatic mode data. When in closed loop, dynamic data is generated by bumping the SP. For model fitting purposes, the controller must be tuned such that the CO takes clear and sudden actions in response to the SP changes, and these must force PV movements that dominate the measurement noise.

Because a closed loop approach makes it possible to generate integrating process dynamic test data that begins at steady state, we can use model fitting software much like we did in this closed-loop set point driven study.

We see the pumped tank process under P-Only control. As shown in the left half of the plot, while the major disturbance is quiet and at its design level, we are able to obtain good set point tracking performance with P-Only control.

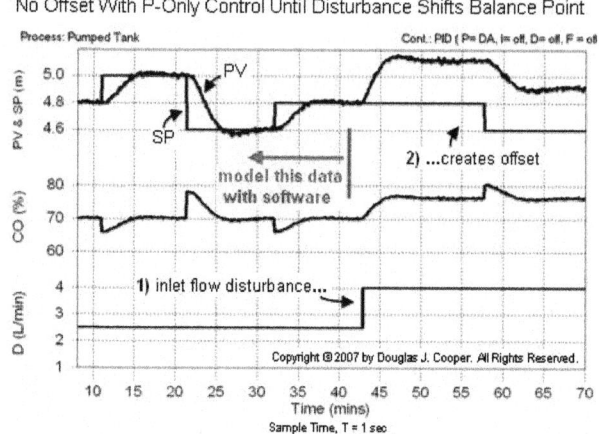

P-Only controller is able to provide good SP tracking performance with no offset as long the major disturbances are quiet and at their design values. Industrial processes can have many disturbances that impact operation. If any one of them changes, as happens in the above plot at roughly 43 minutes, then the simple P-Only controller is incapable of eliminating what becomes a sustained offset (*i.e.*, incapable of making e(t) = SP − PV = 0)

This is why the integral action of a PI controller offers value even though the process itself possesses a naturally integrating behavior.

Note: in a surge tank where exit flow smoothing is more important than maintaining the measured level at SP, offset may not be considered a problem to be solved. Each situation must be considered on its own merits.

The red label in the above plot indicates that the left half contains dynamic response data that begins at steady state and that is not corrupted by disturbance changes.

Step 3: Fit a FOPDT Integrating Model to the Dynamic Process Data

We obtain an approximating description of the closed loop CO to PV dynamic behavior by fitting the process data with a first order plus dead time integrating (FOPDT Integrating) model of the form:

$$\frac{dPV(t)}{dt} = Kp^* \cdot CO(t - \theta p)$$

where

Kp^* = integrator gain, with units [=] PV/(CO ·time)

θp = dead time, with units [=] time

Cropping the data and fitting the FOPDT Integrator model takes but a few mouse clicks with a commercial software tool.

The results of the automated model fit. The visual similarity between the model and data gives us confidence that we have a meaningful description of the dynamic behavior of this process. The Kp* and -p for this approximating model are shown at the bottom of the plot.

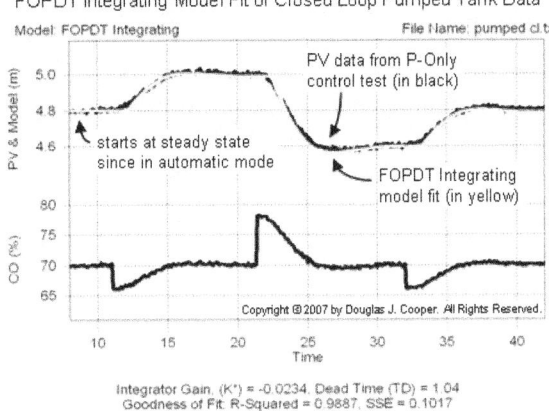

FOPDT Integrating Model Fit of Closed Loop Pumped Tank Data

Integrator Gain, (K*) = -0.0234, Dead Time (TD) = 1.04
Goodness of Fit: R-Squared = 0.9887, SSE = 0.1017

The model fitting software performs a systematic search for the combination of model parameters that minimizes the sum of squared errors (SSE), computed as:

$$SSE = \sum_{i=1}^{N} [\text{Measured PV}_i - \text{Model PV}_i]^2$$

The Measured PV is the actual data collected from our process. The Model PV is computed using the model parameters from the search routine and the actual

CO data from the file. N is the total number of samples in the file. In general, the smaller the SSE, the better the model describes the data.

	- hand fit -	*- software model fits -*	
	Open Loop	**Open Loop**	**P-Only**
Integrator gain, Kp* (m/%·min)	−0.025	−0.023	−0.023
Dead time, θp (min)	1.0	1.0	1.0

Step 4: Use the FOPDT Integrating Parameters to Complete the Design

• *Sample Time,* T

The design and tuning recipe for integrating processes suggests setting the loop sample time, T, at one-tenth the process dead time or faster (*i.e.,* $T \le 0.1\,\theta p$). Faster sampling provides equally good, but not better, performance.

In this study, $T \le 0.1(1.0\text{ min})$, so T should be 6 seconds or less. We meet this with the sample time option available from virtually all commercial vendors:

◊ sample time, T = 1 sec

• *Control Action (Direct/Reverse)*

The pumped tank has a negative Kp*, so when CO increases, PV decreases in response. Since a controller must provide negative feedback, if the process is reverse acting, the controller must be direct acting. Thus, if the PV is too high, the controller must increase the CO to correct the error. Since the controller moves in the same direction as the problem, we specify:

◊ controller is direct acting

• *Computing Controller Error, e(t)*

Set point, SP, is manually entered into a controller. The measured PV comes from the sensor (our wire in). Since SP and PV are known values, then at every loop sample time, T, controller error can be directly computed as:

◊ error, e(t) = SP − PV

• *Determining Bias Value,* CO_{bias}

The lack of a natural balance point with integrating processes makes the determination of a design CO_{bias} problematic. The solution is to use bumpless transfer. That is, when switching to automatic, initialize SP to the current value of PV and CO_{bias} to the current value of CO (most commercial controllers are already programmed this way). By choosing our current operation as our design state at switchover, there is no corrective actions needed by the controller and it can smoothly engage, thus:

◊ controller bias, CO_{bias} = CO that exists at switch over

• *Controller Gain, Kc, and Reset Time, Ti*

We use our FOPDT Integrating model parameters in the industry-proven Internal Model Control (IMC) tuning correlations to compute PI tuning values.

Though all PI forms are equally capable, we use the dependent, ideal form of the PI algorithm in this study:

$$CO = CO_{bias} + Kc \cdot e(t) + \frac{Kc}{Ti} \int e(t)\, dt$$

The first step in using the IMC correlations is to compute Tc, the closed loop time constant. Tc describes how active our controller should be in responding to a set point change or in rejecting a disturbance. *For integrating processes, the design and tuning recipe suggests:*

$$Tc = 3\theta p = 3(1.0 \text{ min}) = 3 \text{ min}$$

With Tc computed, the PI controller gain, Kc, and reset time, Ti, are computed as:

$$Kc = \frac{1}{Kp^*} \frac{2\,Tc + \theta p}{(Tc + \theta p)^2} \qquad Ti = 2Tc + \theta p$$

Substituting the Kp^*, θp *and* Tc identified above into these tuning correlations, we compute:

$$Kc = \frac{1}{-0.023} \frac{2(3) + 1}{(3 + 1)^2} \qquad Ti = 2(3) + 1$$

or

$Kc = -19 \text{ m}/\%$

$Ti = 7 \text{ min}$

Below is the performance of this PI controller (with $Kc = -19$ and $Ti = 7$) on the pumped tank.

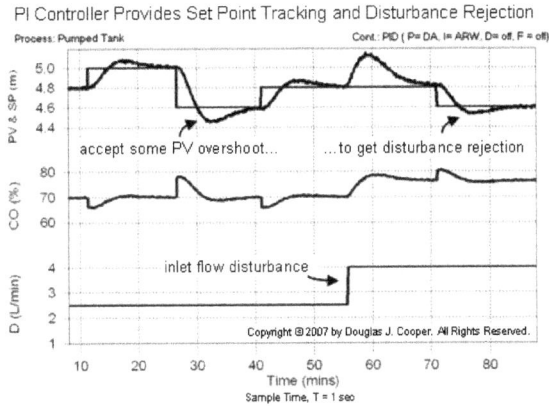

As labeled in the plot, our PI control set point response now includes some overshoot. Recall that the P-Only controller can provide a rapid set point response with no overshoot, that is, until a disturbance changes the balance point of the process.

The benefit of integral action is that when a disturbance occurs, a PI controller can reject the upset and return the PV to set point. This is because the constant summing of integral action continues to move the CO until controller error is driven to zero.

Thus, PI control requires that we accept some overshoot during set point tracking in exchange for the ability to reject disturbances. In many industrial applications, this is considered a fair trade.

Tuning Sensitivity Study

Below is the set point tracking performance of our PI controller on the pumped tank. The controller tuning (Kc = –19 m/%, Ti = 7 min) is determined as detailed in the step-by-step recipe above.

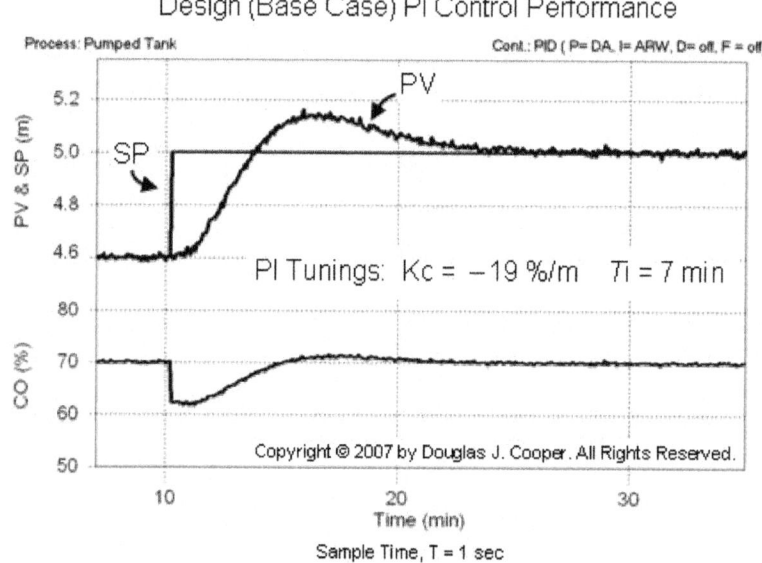

Some questions to consider are:

- How does performance vary as the tuning values change?
- How can we avoid overshoot if we find such behavior undesirable?

Below is a tuning map for a PI controller implemented on the pumped tank integrating process. The center plot is the identical base case performance plot shown above.

The complete map shows set point tracking performance when controller gain (Kc) and reset time (Ti) are individually doubled and halved.

While "good" or "best" performance is a matter best decided by the operations staff, the above map makes it clear that our recipe does an excellent job of meeting the desire for a reasonably rapid rise, a modest overshoot and a quick settling time.

Impact of Kc and T_i on Integrating Process for PI Controller: $CO = CO_{bias} + Kc\,e(t) + \dfrac{Kc}{T_i}\int e(t)\,dt$

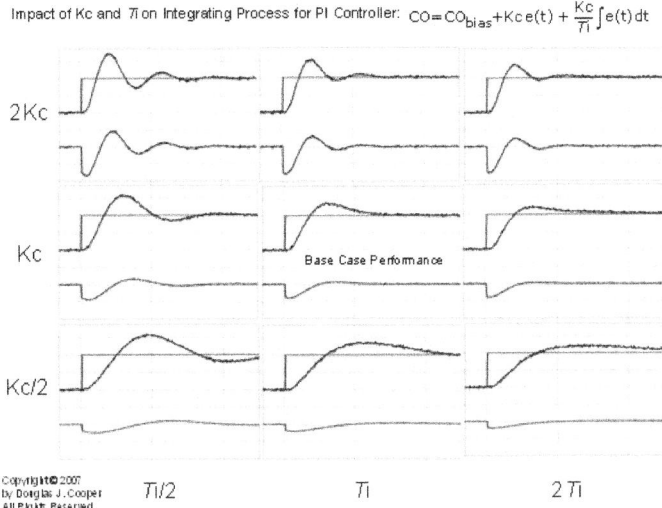

2Kc

Kc

Base Case Performance

Kc/2

Copyright © 2007
by Douglas J. Cooper
All Rights Reserved.

$T_i/2$ T_i $2\,T_i$

Unfortunately, eliminating overshoot altogether does not appear to be one of our options for PI control of integrating processes.

Final Thoughts

The design and tuning recipe for integrating processes provides the above base case performance with minimal testing on our process. In a manufacturing environment where we need a fast solution with minimal disruption, this recipe is certainly one to have in our tool box.

Chapter 3

Automatic Control System

The study and design of automatic Control Systems, a field known as control engineering, has become important in modern technical society. From devices as simple as a toaster or a toilet, to complex machines like space shuttles and power steering, control engineering is a part of our everyday life. This book introduces the field of control engineering and explores some of the more advanced topics in the field. Note, however, that control engineering is a very large field, and this book serves as a foundation of control engineering and introduction to selected advanced topics in the field. Topics in this book are added at the discretion of the authors, and represent the available expertise of our contributors.

Control systems are components that are added to other components, to increase functionality, or to meet a set of design criteria. For example:

We have a particular electric motor that is supposed to turn at a rate of 40 RPM. To achieve this speed, we must supply 10 Volts to the motor terminals. However, with 10 volts supplied to the motor at rest, it takes 30 seconds for our motor to get up to speed. This is valuable time lost.

This simple example, however can be complex to both users and designers of the motor system. It may seem obvious that the motor should start at a higher voltage, so that it accelerates faster. Then we can reduce the supply back down to 10 volts once it reaches ideal speed.

This is clearly a simplistic example, but it illustrates an important point: we can add special "Controller units" to preexisting systems, to improve performance and meet new system specifications.

Control System

A Control System is a device, or a collection of devices that manage the behavior of other devices. Some devices are not controllable. A control system is an interconnection of components connected or related in such a manner as to command, direct, or regulate itself or another system.

Controller

A controller is a control system that manages the behavior of another device or system.

Compensator

A Compensator is a control system that regulates another system, usually by conditioning the input or the output to that system. Compensators are typically employed to correct a single design flaw, with the intention of affecting other aspects of the design in a minimal manner.

There are essentially two methods to approach the problem of designing a new control system: the Classical Approach, and the Modern Approach.

CLASSICAL AND MODERN

Classical and Modern control methodologies are named in a misleading way, because the group of techniques called "Classical" were actually developed later than the techniques labeled "Modern". However, in terms of developing control systems, Modern methods have been used to great effect more recently, while the Classical methods have been gradually falling out of favor. Most recently, it has been shown that Classical and Modern methods can be combined to highlight their respective strengths and weaknesses.

Classical Methods, which this book will consider first, are methods involving the Laplace Transform domain. Physical systems are modeled in the so-called "time domain", where the response of a given system is a function of the various inputs, the previous system values, and time. As time progresses, the state of the system and its response change. However, time-domain models for systems are frequently modeled using high-order differential equations which can become impossibly difficult for humans to solve and some of which can even become impossible for modern computer systems to solve efficiently. To counteract this problem, integral transforms, such as the Laplace Transform and the Fourier Transform, can be employed to change an Ordinary Differential Equation (ODE) in the time domain into a regular algebraic polynomial in the transform domain. Once a given system has been converted into the transform domain it can be manipulated with greater ease and analyzed quickly by humans and computers alike.

Modern Control Methods, instead of changing domains to avoid the complexities of time-domain ODE mathematics, converts the differential equations into a system of lower-order time domain equations called State Equations, which can then be manipulated using techniques from linear algebra. This book will consider Modern Methods second.

A third distinction that is frequently made in the realm of control systems is to divide analog methods (classical and modern, described above) from digital methods. Digital Control Methods were designed to try and incorporate the emerging power of computer systems into previous control methodologies. A special transform, known as the Z-Transform, was developed that can adequately describe digital systems, but at the same time can be converted (with some effort) into the Laplace domain. Once in the Laplace domain, the digital system can be manipulated and analyzed in a very similar manner to Classical analog systems. For this reason, this book will not make a hard and fast distinction between Analog and Digital systems, and instead will attempt to study both paradigms in parallel.

Who is This Book For?

This book is intended to accompany a course of study in under-graduate and graduate engineering. As has been mentioned previously, this book is not focused on any particular discipline within engineering, however any person who wants to make use of this material should have some basic background in the Laplace transform (if not other transforms), calculus, *etc.* The material in this book may be used to accompany several semesters of study, depending on the program of your particular college or university. The study of control systems is generally a topic that is reserved for students in their 3rd or 4th year of a 4 year undergraduate program, because it requires so much previous information. Some of the more advanced topics may not be covered until later in a graduate program.

Many colleges and universities only offer one or two classes specifically about control systems at the undergraduate level. Some universities, however, do offer more than that, depending on how the material is broken up, and how much depth that is to be covered. Also, many institutions will offer a handful of graduate-level courses on the subject. This book will attempt to cover the topic of control systems from both a graduate and undergraduate level, with the advanced topics built on the basic topics in a way that is intuitive. As such, students should be able to begin reading this book in any place that seems an appropriate starting point, and should be able to finish reading where further information is no longer needed.

What are the Prerequisites?

Understanding of the material in this book will require a solid mathematical foundation. This book does not currently explain, nor will it ever try to fully explain most of the necessary mathematical tools used in this text. For that reason, the reader is expected to have read the following wikibooks, or have background knowledge comparable to them:

Algebra

Calculus

The reader should have a good understanding of differentiation and integration. Partial differentiation, multiple integration, and functions of multiple variables will be used occasionally, but the students are not necessarily required to know those subjects well. These advanced calculus topics could better be treated as a co-requisite instead of a pre-requisite.

Linear Algebra

State-space system representation draws heavily on linear algebra techniques. Students should know how to operate on matrices. Students should understand basic matrix operations (addition, multiplication, determinant, inverse, transpose). Students would also benefit from a prior understanding of Eigenvalues and Eigenvectors, but those subjects are covered in this text.

Ordinary Differential Equations

All linear systems can be described by a linear ordinary differential equation. It is beneficial, therefore, for students to understand these equations. Much of this book describes methods to analyze these equations. Students should know what a differential equation is, and they should also know how to find the general solutions of first and second order ODEs.

How is this Book Organized?

This book will be organized following a particular progression. The basics of system theory, and it will offer a brief refresher on integral transforms. The contain a brief primer on digital information, for students who are not necessarily familiar with them. This is done so that digital and analog signals can be considered in parallel throughout the rest of the book. Next, this book will introduce the state-space method of system description and control. The state-space and transform methods interchangeably (and occasionally simultaneously).

As the subject matter of this book expands, so too will the prerequisites. For instance, when this book is expanded to cover nonlinear systems, a basic background knowledge of nonlinear mathematics will be required.

Versions

Each different version is composed of the chapters of this book, included in a different order. This book covers a wide range of information, so if you don't need all the information that this book has to offer, perhaps one of the other versions would be right for you and your educational needs.

Each separate version has a table of contents outlining the different chapters that are included in that version. Also, each separate version comes complete with a printable version, and some even come with PDF versions as well.

Differential Equations Review

Implicit in the study of control systems is the underlying use of differential equations. Even if they aren't visible on the surface, all of the continuous-time systems that we will be looking at are described in the time domain by ordinary differential equations (ODE), some of which are relatively high-order.

Let's review some differential equation basics. Consider the topic of interest from a bank. The amount of interest accrued on a given principal balance (the amount of money you put into the bank) P, is given by:

$$\frac{dP}{dt} = rP$$

Where $\frac{dP}{dt}$ is the interest (rate of change of the principal), and r is the interest rate. Notice in this case that P is a function of time (t), and can be rewritten to reflect that:

$$\frac{dP(t)}{dt} = rP(t)$$

To solve this basic, first-order equation, we can use a technique called "separation of variables", where we move all instances of the letter P to one side, and all instances of t to the other:

$$\frac{dP(t)}{P(t)} = r \, dt$$

And integrating both sides gives us:

$$\ln|P(t)| = rt + C$$

This is all fine and good, but generally, we like to get rid of the logarithm, by raising both sides to a power of e:

$$P(t) = e^{rt+C}$$

Where we can separate out the constant as such:

$$D = e^{C}$$
$$P(t) = De^{rt}$$

D is a constant that represents the initial conditions of the system, in this case the starting principal.

Differential equations are particularly difficult to manipulate, especially once we get to higher-orders of equations. Luckily, several methods of abstraction have been created that allow us to work with ODEs, but at the same time, not have to worry about the complexities of them. The classical method, as described above, uses the Laplace, Fourier, and Z Transforms to convert ODEs in the time domain into polynomials in a complex domain. These complex polynomials are significantly easier to solve than the ODE counterparts. The Modern method instead breaks differential equations into systems of low-order equations, and expresses this system in terms of matrices. It is a common precept in ODE theory that an ODE of order N can be broken down into N equations of order 1.

Readers who are unfamiliar with differential equations might be able to read and understand the material in this book reasonably well.

History

The field of control systems started essentially in the ancient world. Early civilizations, notably the Greeks and the Arabs were heavily preoccupied with the accurate measurement of time, the result of which were several "water clocks" that were designed and implemented.

However, there was very little in the way of actual progress made in the field of engineering until the beginning of the renaissance in Europe. Leonhard Euler (for whom Euler's Formula is named) discovered a powerful integral transform, but Pierre-Simon Laplace used the transform (later called the Laplace Transform) to solve complex problems in probability theory.

Joseph Fourier was a court mathematician in France under Napoleon I. He created a special function decomposition called the Fourier Series, that was later generalized into an integral transform, and named in his honor (the Fourier Transform).

The "golden age" of control engineering occurred between 1910-1945, where mass communication methods were being created and two world wars were being fought. During this period, some of the most famous names in controls engineering were doing their work: Nyquist and Bode.

Hendrik Wade Bode and Harry Nyquist, especially in the 1930's while working with Bell Laboratories, created the bulk of what we now call "Classical Control Methods". These methods were based off the results of the Laplace and Fourier Transforms, which had been previously known, but were made popular by Oliver Heaviside around the turn of the century. Previous to Heaviside, the transforms were not widely used, nor respected mathematical tools.

Bode is credited with the "discovery" of the closed-loop feedback system, and the logarithmic plotting technique that still bears his name (bode plots). Harry Nyquist did extensive research in the field of system stability and information theory. He created a powerful stability criteria that has been named for him (The Nyquist Criteria).

Modern control methods were introduced in the early 1950's, as a way to by-pass some of the shortcomings of the classical methods. Rudolf Kalman is famous for his work in modern control theory, and an adaptive controller called the Kalman Filter was named in his honor. Modern control methods became increasingly popular after 1957 with the invention of the computer, and the start of the space program. Computers created the need for digital control methodologies, and the space program required the creation of some "advanced" control techniques, such as "optimal control", "robust control", and "nonlinear control". These last subjects, and several more, are still active areas of study among research engineers.

Branches of Control Engineering

Here we are going to give a brief listing of the various different methodologies within the sphere of control engineering. Oftentimes, the lines between these methodologies are blurred, or even erased completely.

Classical Controls

Control methodologies where the ODEs that describe a system are transformed using the Laplace, Fourier, or Z Transforms, and manipulated in the transform domain.

Modern Controls

Methods where high-order differential equations are broken into a system of first-order equations. The input, output, and internal states of the system are described by vectors called "state variables".

Robust Control

Control methodologies where arbitrary outside noise/disturbances are accounted for, as well as internal inaccuracies caused by the heat of the system itself, and the environment.

Optimal Control

In a system, performance metrics are identified, and arranged into a "cost function". The cost function is minimized to create an operational system with the lowest cost.

Adaptive Control

In adaptive control, the control changes its response characteristics over time to better control the system.

Nonlinear Control

The youngest branch of control engineering, nonlinear control encompasses systems that cannot be described by linear equations or ODEs, and for which there is often very little supporting theory available.

Game Theory

Game Theory is a close relative of control theory, and especially robust control and optimal control theories. In game theory, the external disturbances are not considered to be random noise processes, but instead are considered to be "opponents". Each player has a cost function that they attempt to minimize, and that their opponents attempt to maximize.

This book will definitely cover the first two branches, and will hopefully be expanded to cover some of the later branches, if time allows.

MATLAB

MATLAB ® is a programming tool that is commonly used in the field of control engineering. MATLAB will not appear in discussions outside these specific sections, although MATLAB may be used in some example problems.

Nearly all textbooks on the subject of control systems, linear systems, and system analysis will use MATLAB as an integral part of the text. Students who are learning this subject at an accredited university will certainly have seen this material in their textbooks, and are likely to have had MATLAB work as part of their classes.

About Formatting

This book will use some simple conventions throughout.

Mathematical Conventions

Mathematical equations will be labeled with the {{eqn}} template, to give them names. Equations that are labeled in such a manner are important, and should be taken special note of. For instance, notice the label to the right of this equation:

$$f(t) = \mathcal{L}^{-1}\left\{F(s)\right\} = \frac{1}{2\pi i}\int_{c-i\infty}^{c+i\infty} e^{st}(s)\,ds$$

Equations that are named in this manner will also be copied into the List of Equations Glossary in the end of the book, for an easy reference.

Italics will be used for English variables, functions, and equations that appear in the main text. For example e, j, $f(t)$ and $X(s)$ are all italicized. It contains a LaTeX mathematics formatting engine, although an attempt will be made not to employ formatted mathematical equations inline with other text because of the difference in size and font. Greek letters, and other non-English characters will not be italicized in the text unless they appear in the midst of multiple variables which are italicized (as a convenience to the editor).

Scalar time-domain functions and variables will be denoted with lower-case letters, along with a t in parenthesis, such as: $x(t)$, $y(t)$, and $h(t)$. Discrete-time functions will be written in a similar manner, except with an $[n]$ instead of a (t).

Fourier, Laplace, Z, and Star transformed functions will be denoted with capital letters followed by the appropriate variable in parenthesis. For example: $F(s)$, $X(j\omega)$, $Y(z)$, and $F^*(s)$.

Matrices will be denoted with capital letters. Matrices which are functions of time will be denoted with a capital letter followed by a t in parenthesis. For example: $A(t)$ is a matrix, $a(t)$ is a scalar function of time.

Transforms of time-variant matrices will be displayed in uppercase bold letters, such as H(s).

Math equations rendered using LaTeX will appear on separate lines, and will be indented from the rest of the text.

SYSTEM IDENTIFICATION

Systems

Systems, in one sense, are devices that take input and produce an output. A system can be thought to operate on the input to produce the output. The output is related to the input by a certain relationship known as the system response. The system response usually can be modeled with a mathematical relationship between the system input and the system output.

System Properties

Physical systems can be divided up into a number of different categories, depending on particular properties that the system exhibits. Some of these system classifications are very easy to work with and have a large theory base for analysis. Some system classifications are very complex and have still not been investigated with any degree of success. By properly identifying the properties of a system, certain analysis and design tools can be selected for use with the system.

The focus primarily on linear time-invariant (LTI) systems. LTI systems are the easiest class of system to work with, and have a number of properties that make them ideal to study.

An introduction to system identification and least squares techniques can be found here. An introduction to parameter identification techniques can be found here.

Initial Time

The initial time of a system is the time before which there is no input. Typically, the initial time of a system is defined to be zero, which will simplify the analysis significantly. Some techniques, such as the Laplace Transform require that the initial time of the system be zero. The initial time of a system is typically denoted by t_0.

The value of any variable at the initial time t_0 will be denoted with a 0 subscript. For instance, the value of variable x at time t_0 is given by:

$$x(t_0) = x_0$$

Likewise, any time t with a positive subscript are points in time *after t_0*, in ascending order:

$$t_0 \le t_1 \le t_2 \le \cdots \le t_n$$

So t_1 occurs after t_0, and t_2 occurs after both points. In a similar fashion above, a variable with a positive subscript (unless specifying an index into a vector) also occurs at that point in time:

$$x(t_1) = x_1$$
$$x(t_2) = x_2$$

This is valid for all points in time t.

Additivity

A system satisfies the property of additivity, if a sum of inputs results in a sum of outputs. By definition: an input of $x_3(t) = x_1(t) + x_2(t)$ results in an output of $y_3(t) = y_1(t) + y_2(t)$. To determine whether a system is additive, use the following test:

Given a system f that takes an input x and outputs a value y, assume two inputs (x_1 and x_2) produce two outputs:

$$y_1 = f(x_1)$$
$$y_2 = f(x_2)$$

Now, create a composite input that is the sum of the previous inputs:

$$x_3 = x_1 + x_2$$

Then the system is additive if the following equation is true:

$$y_3 - f(x_3) = f(x_1 + x_2) = f(x_1) + f(x_2) = y_1 + y_2$$

Systems that satisfy this property are called additive. Additive systems are useful because a sum of simple inputs can be used to analyze the system response to a more complex input.

Example: Sinusoids

Given the following equation:

$$y(t) = \sin(3x(t))$$

Create a sum of inputs as:

$$x(t) = x_1(t) + x_2(t)$$

and construct the expected sum of outputs:

$$y(t) = y_1(t) + y_2(t)$$

Now, substituting these values into our equation, test for equality:

$$y_1(t) + y_2(t) = \sin(3[x_1(t) + x_2(t)])$$

The equality is not satisfied, and therefore the sine operation is not additive.

Homogeneity

A system satisfies the condition of homogeneity if an input scaled by a certain factor produces an output scaled by that same factor. By definition: an input of

ax_1 results in an output of ay_1. In other words, to see if function $f()$ is homogeneous, perform the following test:

Stimulate the system f with an arbitrary input x to produce an output y:

$$y = f(x)$$

Now, create a second input x_1, scale it by a multiplicative factor C (C is an arbitrary constant value), and produce a corresponding output y_1:

$$y_1 = f(Cx_1)$$

Now, assign x to be equal to x_1:

$$x_1 = x$$

Then, for the system to be homogeneous, the following equation must be true:

$$y_1 = f(Cx) = Cf(x) = Cy$$

Systems that are homogeneous are useful in many applications, especially applications with gain or amplification.

Example: Straight-Line

Given the equation for a straight line:

$$y = f(x) = 2x + 3$$
$$y_1 = f(Cx_1) = 2\,(Cfx_1) + 3 = C2x_1 + 3$$
$$x_1 = x$$

Comparing the two results, it is easy to see they are not equal:

$$y_1 = C2x + 3 \neq Cf = C(2x + 3) = C2x + C3$$

Therefore, the equation is not homogeneous.

Linearity

A system is considered linear if it satisfies the conditions of Additivity and Homogeneity. In short, a system is linear if the following is true:

Take two arbitrary inputs, and produce two arbitrary outputs:

$$y_1 = f(x_1)$$
$$y_2 = f(x_2)$$

Now, a linear combination of the inputs should produce a linear combination of the outputs:

$$f(Ax + By) = f(Ax) + f(By) = Af(x) + Bf(y)$$

This condition of additivity and homogeneity is called superposition. A system is linear if it satisfies the condition of superposition.

Example: Linear Differential Equations

Is the following equation linear:

$$\frac{dy(t)}{dt} + y(t) = x(t)$$

To determine whether this system is linear, construct a new composite input:

$$x(t) = Ax_1(t) + Bx_2(t)$$

Now, create the expected composite output:

$$y(t) = Ay_1(t) + By_2(t)$$

Substituting the two into our original equation:

$$\frac{d[Ay_1(t) + By_2(t)]}{dt} + [Ay_1(t) + By_2(t)] = Ax_1(t) + Bx_2(t)$$

Factor out the derivative operator, as such:

$$\frac{d}{dt}[Ay_1(t) + By_2(t)] + [Ay_1(t) + By_2(t)] = Ax_1(t) + Bx_2(t)$$

Finally, convert the various composite terms into the respective variables, to prove that this system is linear:

$$\frac{dy(t)}{dt} + y(t) = x(t)$$

For the record, derivatives and integrals are linear operators, and ordinary differential equations typically are linear equations.

Memory

A system is said to have memory if the output from the system is dependent on past inputs (or future inputs!) to the system. A system is called memoryless if the output is only dependent on the current input. Memoryless systems are easier to work with, but systems with memory are more common in digital signal processing applications.

Systems that have memory are called dynamic systems, and systems that do not have memory are static systems.

Causality

Causality is a property that is very similar to memory. A system is called causal if it is only dependent on past and/or current inputs. A system is called anti-causal if the output of the system is dependent only on future inputs. A system is called non-causal if the output depends on past and/or current and future inputs.

A system design that is not causal cannot be physically implemented. If the system can't be built, the design is generally worthless.

Time-Invariance

A system is called time-invariant if the system relationship between the input and output signals is not dependent on the passage of time. If the input signal $x(t)$ produces an output $y(t)$ then any time shifted input, $x(t + \delta)$, results in a time-shifted output $y(t + \delta)$. This property can be satisfied if the transfer function of the system is not a function of time except expressed by the input and output. If a system is time-invariant then the system block is commutative with an arbitrary delay.

To determine if a system f is time-invariant, perform the following test:

Apply an arbitrary input x to a system and produce an arbitrary output y :

$$y(t) = f(x(t))$$

Apply a second input x_1 to the system, and produce a second output:

$$y_1(t) = f(x_1(t))$$

Now, assign x_1 to be equal to the first input x, time-shifted by a given constant value δ :

$$x_1(t) = x(t - \delta)$$

Finally, a system is time-invariant if y_1 is equal to y shifted by the same value δ :

$$y_1(t) = y(t - \delta)$$

LTI Systems

A system is considered to be a Linear Time-Invariant (LTI) system if it satisfies the requirements of time-invariance and linearity.

Systems which are not LTI are more common in practice, but are much more difficult to analyze.

Lumpedness

A system is said to be lumped if one of the two following conditions are satisfied:

1. There are a finite number of states that the system can be in.
2. There are a finite number of state variables.

The concept of "states" and "state variables" are relatively advanced, and they will be discussion about modern controls.

Systems which are not lumped are called distributed. A simple example of a distributed system is a system with delay, that is,

$$A(s)\, y(t) = B(s)\, u(t - \tau),$$

which has an infinite number of state variables. However, although distributed systems are quite common, they are very difficult to analyze in practice, and there are few tools available to work with such systems. Fortunately, in most cases, a delay can be sufficiently modeled with the Pade approximation.

Relaxed

A system is said to be relaxed if the system is causal, and at the initial time t_0 the output of the system is zero, *i.e.*, there is no stored energy in the system.

$$y(t_0) = f(x(t_0)) = 0$$

In terms of differential equations, a relaxed system is said to have "zero initial state". Systems without an initial state are easier to work with, but systems that are not relaxed can frequently be modified to approximate relaxed systems.

Stability

Control Systems engineers will frequently say that an unstable system has "exploded". Some physical systems actually can rupture or explode when they go unstable.

Stability is a very important concept in systems, but it is also one of the hardest function properties to prove. There are several different criteria for system stability, but the most common requirement is that the system must produce a finite output when subjected to a finite input. For instance, if 5 volts is applied to the input terminals of a given circuit, it would be best if the circuit output didn't approach infinity, and the circuit itself didn't melt or explode. This type of stability is often known as "Bounded Input, Bounded Output" stability, or BIBO.

There are a number of other types of stability, most of which are based off the concept of BIBO stability.

Inputs and Outputs

Systems can also be categorized by the number of inputs and the number of outputs the system has. Consider a television as a system, for instance. The system has two inputs: the power wire and the signal cable. It has one output: the video display. A system with one input and one output is called single-input, single output, or SISO. a system with multiple inputs and multiple outputs is called multi-input, multi-output, or MIMO.

DIGITAL AND ANALOG

Digital and Analog

There is a significant distinction between an analog system and a digital system, in the same way that there is a significant difference between analog and digital data. This book is going to consider both analog and digital topics, so it is worth taking some time to discuss the differences, and to display the different notations that will be used with each.

Continuous Time

A signal is called continuous-time if it is defined at every time t.

A system is a continuous-time system if it takes a continuous-time input signal, and outputs a continuous-time output signal. Here is an example of an analog waveform:

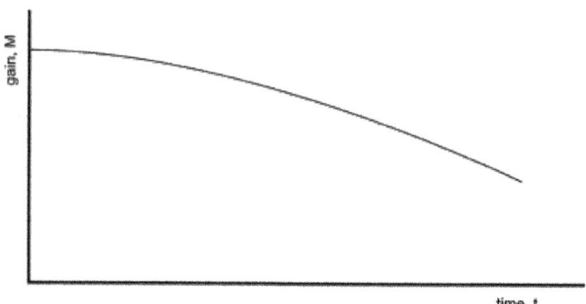

Discrete Time

A signal is called discrete-time if it is only defined for particular points in time. A discrete-time system takes discrete-time input signals, and produces discrete-time output signals. The following image shows the difference between an analog waveform and the sampled discrete time equivalent:

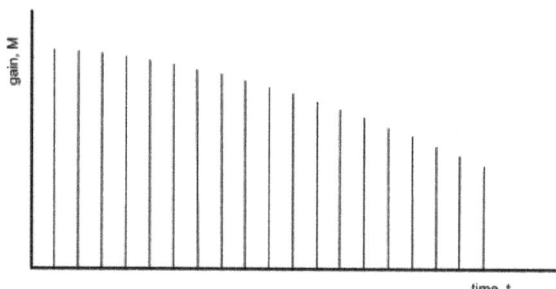

Quantized

A signal is called Quantized if it can only be certain values, and cannot be other values. This concept is best illustrated with examples:

1. Students with a strong background in physics will recognize this concept as being the root word in "Quantum Mechanics". In quantum mechanics, it is known that energy comes only in discrete packets. An electron bound to an atom, for example, may occupy one of several discrete energy levels, but not intermediate levels.

2. Another common example is population statistics. For instance, a common statistic is that a household in a particular country may have an average of "3.5 children", or some other fractional number. Actual households may have 3 children, or they may have 4 children, but no household has 3.5 children.

3. People with a computer science background will recognize that integer variables are quantized because they can only hold certain integer values, not fractions or decimal points.

The last example concerning computers is the most relevant, because quantized systems are frequently computer-based. Systems that are implemented with computer software and hardware will typically be quantized.

Here is an example waveform of a quantized signal. Notice how the magnitude of the wave can only take certain values, and that creates a step-like appearance. This image is discrete in magnitude, but is continuous in time:

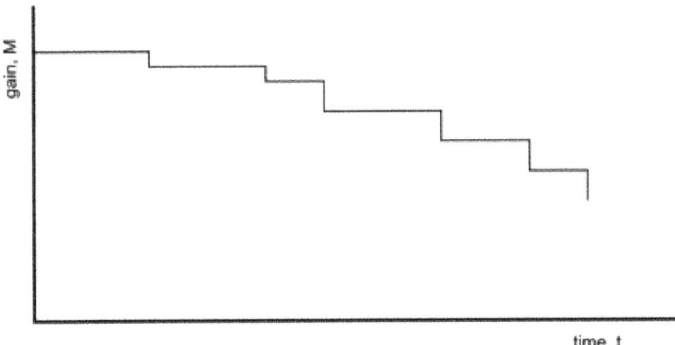

Analog

By definition:

Analog

A signal is considered analog if it is defined for all points in time and if it can take any real magnitude value within its range.

An analog system is a system that represents data using a direct conversion from one form to another. In other words, an analog system is a system that is continuous in both time and magnitude.

Example: Motor

If we have a given motor, we can show that the output of the motor (rotation in units of radians per second, for instance) is a function of the voltage that is input to the motor. We can show the relationship as such:

$$\Theta(v) = f(v)$$

Where Θ is the output in terms of Rad/sec, and $f(v)$ is the motor's conversion function between the input voltage (v) and the output. For any value of v we can calculate out specifically what the rotational speed of the motor should be.

Example: Analog Clock

Consider a standard analog clock, which represents the passage of time though the angular position of the clock hands. We can denote the angular position of the hands of the clock with the system of equations:

$$\phi_h = f_h(t)$$
$$\phi_m = f_m(t)$$
$$\phi_s = f_s(t)$$

Where φ_h is the angular position of the hour hand, φ_m is the angular position of the minute hand, and φ_s is the angular position of the second hand. The positions of all the different hands of the clock are dependent on functions of time.

Different positions on a clock face correspond directly to different times of the day.

Digital

Digital data is represented by discrete number values. By definition:

Digital

A signal or system is considered digital if it is both discrete-time and quantized.

Digital data always have a certain granularity, and therefore there will almost always be an error associated with using such data, especially if we want to account for all real numbers. The tradeoff, of course, to using a digital system is that our powerful computers with our powerful, Moore's law microprocessor units, can be instructed to operate on digital data only. This benefit more than makes up for the shortcomings of a digital representation system.

Discrete systems will be denoted inside square brackets, as is a common notation in texts that deal with discrete values. For instance, we can denote a discrete data set of ascending numbers, starting at 1, with the following notation:

$$x[n] = [1\ 2\ 3\ 4\ 5\ 6...]$$

n, or other letters from the central area of the alphabet (m, i, j, k, l, for instance) are commonly used to denote discrete time values. Analog, or "non-discrete" values are denoted in regular expression syntax, using parenthesis. Here is an example of an analog waveform and the digital equivalent. Notice that the digital waveform is discrete in both time and magnitude:

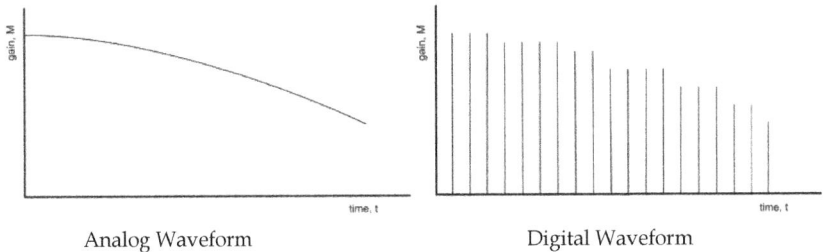

Analog Waveform   Digital Waveform

Example: Digital Clock

As a common example, let's consider a digital clock: The digital clock represents time with binary electrical data signals of 1 and 0. The 1's are usually represented by a positive voltage, and a 0 is generally represented by zero voltage. Counting in binary, we can show that any given time can be represented by a base-2 numbering system:

Minute	Binary Representation
1	1
10	1010
30	11110
59	111011

But what happens if we want to display a fraction of a minute, or a fraction of a second? A typical digital clock has a certain amount of precision, and it cannot express fractional values smaller than that precision.

Hybrid Systems

Hybrid Systems are systems that have both analog and digital components. Devices called samplers are used to convert analog signals into digital signals, and Devices called reconstructors are used to convert digital signals into analog signals. Because of the use of samplers, hybrid systems are frequently called sampled-data systems.

Example: Automobile Computer

Most modern automobiles today have integrated computer systems that monitor certain aspects of the car, and actually help to control the performance of the car. The speed of the car, and the rotational speed of the transmission are analog values, but a sampler converts them into digital values so the car computer can monitor them. The digital computer will then output control signals to other parts of the car, to alter analog systems such as the engine timing, the suspension, the brakes, and other parts. Because the car has both digital and analog components, it is a hybrid system.

Continuous and Discrete

A system is considered continuous-time if the signal exists for all time. Frequently, the terms "analog" and "continuous" will be used interchangeably, although they are not strictly the same.

Discrete systems can come in three flavors:

1. Discrete time (sampled)
2. Discrete magnitude (quantized)
3. Discrete time and magnitude (digital)

Discrete magnitude systems are systems where the signal value can only have certain values. Discrete time systems are systems where signals are only available (or valid) at particular times. Computer systems are discrete in the sense of (3), in that data is only read at specific discrete time intervals, and the data can have only a limited number of discrete values.

A discrete-time system has a sampling time value associated with it, such that each discrete value occurs at multiples of the given sampling time. We will denote the sampling time of a system as T. We can equate the square-brackets notation of a system with the continuous definition of the system as follows:

$$x[n] = x(nT)$$

Notice that the two notations show the same thing, but the first one is typically easier to write, and it shows that the system in question is a discrete system. This book will use the square brackets to denote discrete systems by the sample number n, and parenthesis to denote continuous time functions.

Sampling and Reconstruction

The process of converting analog information into digital data is called "Sampling". The process of converting digital data into an analog signal is called "Reconstruction". We will talk about both processes. Here is an example of a reconstructed waveform. Notice that the reconstructed waveform here is quantized because it is constructed from a digital signal:

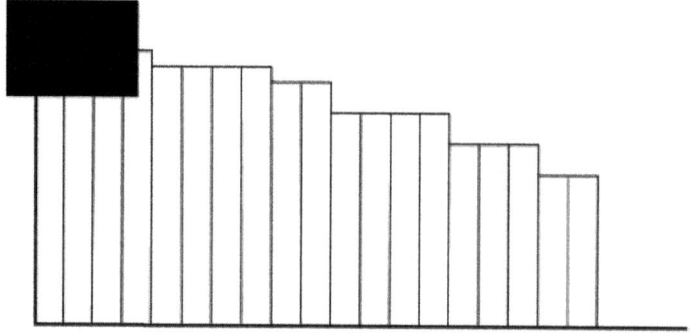

SYSTEM MODELING

The Control Process

It is the job of a control engineer to analyze existing systems, and to design new systems to meet specific needs. Sometimes new systems need to be designed, but more frequently a controller unit needs to be designed to improve the performance of existing systems. When designing a system, or implementing a controller to augment an existing system, we need to follow some basic steps:

1. Model the system mathematically
2. Analyze the mathematical model
3. Design system/controller
4. Implement system/controller and test

The vast majority of this book is going to be focused on (2), the analysis of the mathematical systems.

External Description

An external description of a system relates the system input to the system output without explicitly taking into account the internal workings of the system. The external description of a system is sometimes also referred to as the Input-Output Description of the system, because it only deals with the inputs and the outputs to the system.

$$\xrightarrow{\quad x(t) \quad} \boxed{h(t)} \xrightarrow{\quad y(t) \quad}$$

If the system can be represented by a mathematical function $h(t, r)$, where t is the time that the output is observed, and r is the time that the input is applied. We can relate the system function $h(t, r)$ to the input x and the output y through the use of an integral:

[General System Description]

$$y(t) = \int_{-\infty}^{\infty} h(t,r)\, x\,(r)\, dr$$

This integral form holds for all linear systems, and every linear system can be described by such an equation.

If a system is causal, then there is no output of the system before time r, and we can change the limits of the integration:

$$y(t) = \int_{0}^{t} h(t,r)\, x\,(r)\, dr$$

Time-Invariant Systems

If a system is time-invariant (and causal), we can rewrite the system description equation as follows:

$$y(t) = \int_0^t h(t-r)\, x\,(r)\, dr$$

This equation is known as the convolution integral.

Every Linear Time-Invariant (LTI) system can be used with the Laplace Transform, a powerful tool that allows us to convert an equation from the time domain into the S-Domain, where many calculations are easier. Time-variant systems cannot be used with the Laplace Transform.

Internal Description

If a system is linear and lumped, it can also be described using a system of equations known as state-space equations. In state space equations, we use the variable x to represent the internal state of the system. We then use u as the system input, and we continue to use y as the system output. We can write the state space equations as such:

$$x'(t) = A(t)\, x(t) + B(t)\, u(t)$$
$$y(t) = C(t)\, x(t) + D(t)\, u(t)$$

Complex Descriptions

Systems which are LTI and Lumped can also be described using a combination of the state-space equations, and the Laplace Transform. If we take the Laplace Transform of the state equations that we listed above, we can get a set of functions known as the Transfer Matrix Functions.

Representations

To recap, we will prepare a table with the various system properties, and the available methods for describing the system:

Properties	State-Space Equations	Laplace Transform	Transfer Matrix
Linear, Time-Variant, Distributed	no	no	no
Linear, Time-Variant, Lumped	yes	no	no
Linear, Time-Invariant, Distributed	no	yes	no
Linear, Time-Invariant, Lumped	yes	yes	yes

Analysis

Once a system is modeled using one of the representations listed above, the system needs to be analyzed. We can determine the system metrics and then we can compare those metrics to our specification. If our system meets the specifications we are finished with the design process. However if the system does not meet the specifications (as is typically the case), then suitable controllers and compensators need to be designed and added to the system.

Once the controllers and compensators have been designed, the job isn't finished: we need to analyze the new composite system to ensure that the controllers work properly. Also, we need to ensure that the systems are stable: unstable systems can be dangerous.

Frequency Domain

For proposals, early stage designs, and quick turn around analyses a frequency domain model is often superior to a time domain model. Frequency domain models take disturbance PSDs (Power Spectral Densities) directly, use transfer functions directly, and produce output or residual PSDs directly. The answer is a steady-state response. Oftentimes the controller is shooting for 0 so the steady-state response is also the residual error that will be the analysis output or metric for report.

Table : Frequency Domain Model Inputs and Outputs

Input	Model	Output
PSD	Transfer Function	PSD

Brief Overview of the Math

Frequency domain modeling is a matter of determining the impulse response of a system to a random process.

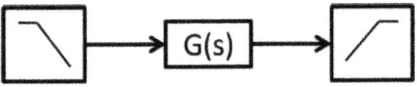

Fig. : Frequency Domain System.

$$S_{YY}(\omega) = G^*(\omega)\, G(\omega)\, S_{XX} = |G(\omega)|\, S_{XX}$$

where

$S_{XX}(\omega)$ is the one-sided input PSD in $\dfrac{magnitude^2}{Hz}$

$G(\omega)$ is the frequency response function of the system and

$S_{YY}(\omega)$ is the one-sided output PSD or auto power spectral density function.

The frequency response function, $G(\omega)$, is related to the impulse response function (transfer function) by

$$g(\tau) = \frac{1}{2\pi} \int_{-\infty}^{\infty} e^{iwt} H(\omega)\, d\omega$$

Note some texts will state that this is only valid for random processes which are stationary. Other texts suggest stationary and ergodic while still others state weakly stationary processes. Some texts do not distinguish between strictly stationary and weakly stationary. From practice, the rule thumb is if the PSD of the input process is the same from hour to hour and day to day then the input PSD can be used and the above equation is valid.

Chapter 4

MODERN CONTROL METHODS

STATE-SPACE EQUATIONS

Time-Domain Approach

The "Classical" method of controls (what we have been studying so far) has been based mostly in the transform domain. When we want to control the system in general we use the Laplace transform (Z-Transform for digital systems) to represent the system, and when we want to examine the frequency characteristics of a system, we use the Fourier Transform. The question arises, why do we do this?

Let's look at a basic second-order Laplace Transform transfer function:

$$\frac{Y(s)}{X(s)} = G(s) = \frac{1+s}{1+2s+5s^2}$$

And we can decompose this equation in terms of the system inputs and outputs:

$$(1+2s+5s^2)Y(s) = (1+s)X(s)$$

Now, when we take the inverse Laplace transform of our equation, we can see that:

$$y(t) + 2\frac{dy(t)}{dt} + 5\frac{d^2y(t)}{dt^2} = x(t) + \frac{dx(t)}{dt}$$

The Laplace transform is transforming the fact that we are dealing with second-order differential equations. The Laplace transform moves a system out of the time-domain into the complex frequency domain, to study and manipulate our systems as algebraic polynomials instead of linear ODEs. Given the complexity of differential equations, why would we ever want to work in the time domain?

It turns out that to decompose our higher-order differential equations into multiple first-order equations, one can find a new method for easily manipulating the system *without having to use integral transforms*. The solution to this problem is state variables . By taking our multiple first-order differential equations, and analyzing them in vector form, we can not only do the same things we were doing in the time domain using simple matrix algebra, but now we can easily account for systems with multiple inputs and multiple outputs, without adding much unnecessary complexity. This demonstrates why the "modern" state-space approach to controls has become popular.

State-Space

In a state space system, the internal state of the system is explicitly accounted for by an equation known as the state equation. The system output is given in terms of a combination of the current system state, and the current system input, through the output equation. These two equations form a system of equations known collectively as state-space equations. The state-space is the vector space that consists of all the possible internal states of the system.

For a system to be modeled using the state-space method, the system must meet this requirement:

The System Must Be "Lumped"

"Lumped" in this context, means that we can find a *finite*-dimensional state-space vector which fully characterises all such internal states of the system.

This text mostly considers linear state space systems, where the state and output equations satisfy the superposition principle and the state space is linear. However, the state-space approach is equally valid for nonlinear systems although some specific methods are not applicable to nonlinear systems.

State

Central to the state-space notation is the idea of a state. A state of a system is the current value of internal elements of the system, that change separately (but not completely unrelated) to the output of the system. In essence, the state of a system is an explicit account of the values of the internal system components. Here are some examples:

Consider an electric circuit with both an input and an output terminal. This circuit may contain any number of inductors and capacitors. The state variables may represent the magnetic and electric fields of the inductors and capacitors, respectively.

Consider a spring-mass-dashpot system. The state variables may represent the compression of the spring, or the acceleration at the dashpot.

Consider a chemical reaction where certain reagents are poured into a mixing container, and the output is the amount of the chemical product produced over time. The state variables may represent the amounts of un-reacted chemicals in the container, or other properties such as the quantity of thermal energy in the container (that can serve to facilitate the reaction).

State Variables

When modeling a system using a state-space equation, we first need to define three vectors:

Input Variables

A SISO (Single Input Single Output) system will only have a single input value, but a MIMO system may have multiple inputs. We need to define all the inputs to the system, and we need to arrange them into a vector.

Output variables

This is the system output value, and in the case of MIMO systems, we may have several. Output variables should be independent of one another, and only dependent on a linear combination of the input vector and the state vector.

State Variables

The state variables represent values from inside the system, that can change over time. In an electric circuit, for instance, the node voltages or the mesh currents can be state variables. In a mechanical system, the forces applied by springs, gravity, and dashpots can be state variables.

We denote the input variables with u, the output variables with y, and the state variables with x. In essence, we have the following relationship:

$$y = f(x, u)$$

Where $f(x, u)$ is our system. Also, the state variables can change with respect to the current state and the system input:

$$x' = g(x, u)$$

Where x' is the rate of change of the state variables.

Multi-Input, Multi-Output

In the Laplace domain, if we want to account for systems with multiple inputs and multiple outputs, we are going to need to rely on the principle of superposition to create a system of simultaneous Laplace equations for each output and each input. For such systems, the classical approach not only doesn't simplify the situation, but because the systems of equations need to be transformed into the frequency domain first, manipulated, and then transformed back into the time domain, they can actually be more difficult to work with.

State-Space Equations

In a state-space system representation, we have a system of two equations: an equation for determining the state of the system, and another equation for determining the output of the system. We will use the variable $y(t)$ as the output of the system, $x(t)$ as the state of the system, and $u(t)$ as the input of the system. We use the notation $x'(t)$ (note the prime) for the first derivative of the state vector of the system, as dependent on the current state of the system and the current input. Symbolically, we say that there are transforms g and h, that display this relationship:

$$x'(t) = g\,[t_0,\, t,\, x(t),\, x(0),\, u(t)]$$
$$y(t) = h\,[t,\, x(t),\, u(t)]$$

The first equation shows that the system state change is dependent on the previous system state, the initial state of the system, the time, and the system inputs. The second equation shows that the system output is dependent on the current system state, the system input, and the current time.

If the system state change $x'(t)$ and the system output $y(t)$ are linear combinations of the system state and input vectors, then we can say the systems are linear systems, and we can rewrite them in matrix form:

[State Equation]

$$x' = A(t)\,x(t) + B(t)\,u(t)$$

[Output Equation]

$$y(t) = C(t)\,x(t) + D(t)\,u(t)$$

If the systems themselves are time-invariant, we can re-write this as follows:

$$x' = Ax(t) + Bu(t)$$
$$y(t) = Cx(t) + Du(t)$$

The State Equation shows the relationship between the system's current state and its input, and the future state of the system. The Output Equation shows the relationship between the system state and its input, and the output. These equations show that in a given system, the current output is dependent on the current input and the current state. The future state is also dependent on the current state and the current input.

It is important to note at this point that the state space equations of a particular system are not unique, and there are an infinite number of ways to represent these equations by manipulating the A, B, C and D matrices using row operations. There are a number of "standard forms" for these matrices, however, that make certain computations easier. Converting between these forms will require knowledge of linear algebra.

State-Space Basis Theorem

Any system that can be described by a finite number of nth order differential equations or n^{th} order difference equations, or any system that can be approximated by them, can be described using state-space equations. The general solutions to the state-space equations, therefore, are solutions to all such sets of equations.

Matrices: A B C D

Our system has the form:

$$x'(t) = g\,[t_0,\, t,\, x(t),\, x(0),\, u(t)]$$
$$y(t) = h\,[t,\, x(t),\, u(t)]$$

We've bolded several quantities to try and reinforce the fact that they can be vectors, not just scalar quantities. If these systems are time-invariant, we can simplify them by removing the time variables:

$$x'(t) = g\,[x(t),\, x(0),\, u(t)]$$
$$y(t) = h\,[x(t),\, u(t)]$$

Now, if we take the partial derivatives of these functions with respect to the input and the state vector at time t_0, we get our system matrices:

$$A = g_x\,[x(0),\, x(0),\, u(0)]$$
$$B = g_u\,[x(0),\, x(0),\, u(0)]$$
$$C = h_x\,[x(0),\, u(0)]$$
$$D = h_u\,[x(0),\, u(0)]$$

In our time-invariant state space equations, we write these matrices and their relationships as:

$$x'(t) = Ax(t) + Bu(t)$$
$$y(t) = Cx(t) + Du(t)$$

We have four constant matrices: A, B, C, and D. We will explain these matrices below:

Matrix A

Matrix A is the system matrix, and relates how the current state affects the state change x'. If the state change is not dependent on the current state, A will be the zero matrix. The exponential of the state matrix, e^{At} is called the state transition matrix.

Matrix B

Matrix B is the control matrix, and determines how the system input affects the state change. If the state change is not dependent on the system input, then B will be the zero matrix.

Matrix C

Matrix C is the output matrix, and determines the relationship between the system state and the system output.

Matrix D

Matrix D is the feed-forward matrix, and allows for the system input to affect the system output directly. A basic feedback system like those we have previously considered do not have a feed-forward element, and therefore for most of the systems we have already considered, the D matrix is the zero matrix.

Matrix Dimensions

Because we are adding and multiplying multiple matrices and vectors together, we need to be absolutely certain that the matrices have compatible dimensions, or else the equations will be undefined. For integer values p, q, and r, the dimensions of the system matrices and vectors are defined as follows:

Vectors	Matrices
• $x : p \times 1$	• A: $p \times p$
• $x' : p \times 1$	• B: $p \times q$
• $u : q \times 1$	• C: $r \times p$
• $y : r \times 1$	• D: $r \times q$

Matrix Dimensions:

$$A: p \times p$$
$$B: p \times q$$
$$C: r \times p$$
$$D: r \times q$$

If the matrix and vector dimensions do not agree with one another, the equations are invalid and the results will be meaningless. Matrices and vectors must have compatible dimensions or they cannot be combined using matrix operations.

For the rest of the book, we will be using the small template on the right as a reminder about the matrix dimensions, so that we can keep a constant notation throughout the book.

Notational Shorthand

The state equations and the output equations of systems can be expressed in terms of matrices A, B, C, and D. Because the form of these equations is always the same, we can use an ordered quadruplet to denote a system. We can use the shorthand (A, B, C, D) to denote a complete state-space representation. Also, because the state equation is very important for our later analyis, we can write an ordered pair (A, B) to refer to the state equation:

$$(A, B) \rightarrow x' = Ax + Bu$$

$$(A, B, C, D) \rightarrow \begin{cases} x' = Ax + Bu \\ y = Cx + Du \end{cases}$$

Obtaining the State-Space Equations

The beauty of state equations, is that they can be used to transparently describe systems that are both continuous and discrete in nature. Some texts will differentiate notation between discrete and continuous cases, but this text will not make such a distinction. Instead we will opt to use the generic coefficient matrices A, B, C and D for both continuous and discrete systems. Occasionally this book may employ the subscript C to denote a continuous-time version of the matrix, and the subscript D to denote the discrete-time version of the same matrix. Other texts may use the letters F, H, and G for continuous systems and Γ, and Θ for use in discrete systems. However, if we keep track of our time-domain system, we don't need to worry about such notations.

From Differential Equations

Let's say that we have a general 3rd order differential equation in terms of input $u(t)$ and output $y(t)$:

$$\frac{d^3y(t)}{dt^3} + a_2\frac{d^2y(t)}{dt^2} + a_1\frac{dy(t)}{dt} + a_0y(t) = u(t)$$

We can create the state variable vector x in the following manner:

$$x_1 = y(t)$$
$$x_2 = \frac{dy(t)}{dt}$$
$$x_3 = \frac{d^2y(t)}{dt^2}$$

Which now leaves us with the following 3 first-order equations:

$$x_1' = x_2$$
$$x_2' = x_3$$
$$x_3' = \frac{d^3y(t)}{dt^3}$$

Now, we can define the state vector x in terms of the individual x components, and we can create the future state vector as well:

$$x = \begin{bmatrix} x_1 \\ x_2 \\ x_3 \end{bmatrix}, x' = \begin{bmatrix} x_1' \\ x_2' \\ x_3' \end{bmatrix}$$

And with that, we can assemble the state-space equations for the system:

$$x' = \begin{bmatrix} 0 & 1 & 0 \\ 0 & 0 & 1 \\ -a_0 & -a_1 & -a_2 \end{bmatrix} x(t) + \begin{bmatrix} 0 \\ 0 \\ 1 \end{bmatrix} u(t)$$

$$y(t) = \begin{bmatrix} 1 & 0 & 0 \end{bmatrix} x(t)$$

Granted, this is only a simple example, but the method should become apparent to most readers.

From Transfer Functions

The method of obtaining the state-space equations from the Laplace domain transfer functions are very similar to the method of obtaining them from the time-domain differential equations. We call the process of converting a system description from the Laplace domain to the state-space domain realization. In general, let's say that we have a transfer function of the form:

$$T(s) = \frac{s^m + a_{m-1}s^{m-1} + \cdots + a_0}{s^n + b_{n-1}s^{n-1} + \cdots + b_0}$$

We can write our A, B, C, and D matrices as follows:

$$A = \begin{bmatrix} 0 & 1 & 0 & \cdots & 0 \\ 0 & 0 & 1 & \cdots & 0 \\ \vdots & \vdots & \vdots & \ddots & \vdots \\ 0 & 0 & 0 & \cdots & 1 \\ -b_0 & -b_1 & -b_2 & \cdots & -b_{n-1} \end{bmatrix}$$

$$B = \begin{bmatrix} 0 \\ 0 \\ \vdots \\ 1 \end{bmatrix}$$

$$C = \begin{bmatrix} a_0 & a_1 & \cdots & a_{m-1} \end{bmatrix}$$

$$D = 0$$

This form of the equations is known as the controllable canonical form of the system matrices.

Notice that to perform this method, the denominator and numerator polynomials must be *monic*, the coefficients of the highest-order term must be 1. If the coefficient of the highest order term is not 1, you must divide your equation by that coefficient to make it 1.

State-Space Representation

As an important note, remember that the state variables x are user-defined and therefore are arbitrary. There are any number of ways to define x for a particular problem, each of which are going to lead to different state space equations.

Note: There are an infinite number of equivalent ways to represent a system using state-space equations. Some ways are better than others. Once these state-space equations are obtained, they can be manipulated to take a particular form if needed.

Consider the previous continuous-time example. We can rewrite the equation in the form

$$\frac{d}{dt}\left[\frac{d^2y(t)}{dt^2} + a_2\frac{dy(t)}{dt} + a_1y(t)\right] + a_0y(t) = u(t)$$

We now define the state variables

$$x_1 = y(t)$$

$$x_2 = \frac{dy(t)}{dt}$$

$$x_3 = \frac{d^2y(t)}{dt^2} + a_2\frac{dy(t)}{dt} + a_1y(t)$$

with first-order derivatives

$$x_1' = \frac{dy(t)}{dt} = x_2$$

$$x_2' = \frac{d^2y(t)}{dt^2} = -a_1x_1 - a_2x_2 + x_3$$

$$x_3' = -a_0y(t) + u(t)$$

The state-space equations for the system will then be given by

$$x' = \begin{bmatrix} 0 & 1 & 0 \\ -a_1 & -a_2 & 1 \\ -a_0 & 0 & 0 \end{bmatrix} x(t) + \begin{bmatrix} 0 \\ 0 \\ 1 \end{bmatrix} u(t)$$

$$y(t) = \begin{bmatrix} 1 & 0 & 0 \end{bmatrix} x(t)$$

x may also be used in any number of variable transformations, as a matter of mathematical convenience. However, the variables y and u correspond to physical signals, and may not be arbitrarily selected, redefined, or transformed as x can be.

Example: Dummy Variables

The attitude control of a particular manned aircraft can be given by:

$$\theta''(t) = \alpha + \delta$$

Where a is the direction the aircraft is traveling in, θ is the direction the aircraft is facing (the attitude), and δ is the angle of the ailerons (the control input from the pilot). This equation is not in a proper format, so we need to produce some dummy-variables:

$$\theta_1 = \theta$$
$$\theta_1' = \theta_2$$
$$\theta_2' = \alpha + \delta$$

This in turn will provide us with our state equation:

$$\begin{bmatrix} \theta_1 \\ \theta_2 \end{bmatrix}' = \begin{bmatrix} 0 & 1 \\ 0 & 0 \end{bmatrix} \begin{bmatrix} \theta_1 \\ \theta_2 \end{bmatrix} + \begin{bmatrix} 0 & 0 \\ 1 & 1 \end{bmatrix} \begin{bmatrix} \alpha \\ \delta \end{bmatrix}$$

As we can see from this equation, even though we have a valid state-equation, the variables θ_1 and θ_2 don't necessarily correspond to any measurable physical event, but are instead dummy variables constructed by the user to help define the system. Note, however, that the variables α and δ do correspond to physical values, and cannot be changed.

Discretization

If we have a system (A, B, C, D) that is defined in continuous time, we can discretize the system so that an equivalent process can be performed using a digital computer. We can use the definition of the derivative, as such:

$$x'(t) = \lim_{T \to 0} \frac{x(t+T) - x(t)}{T}$$

And substituting this into the state equation with some approximation (and ignoring the limit for now) gives us:

$$\lim_{T \to 0} \frac{x(t+T) - x(t)}{T} = Ax(t) + Bu(t)$$
$$x(t+T) = x(t) + Ax(t)T + Bu(t)T$$
$$x(t+T) = (1 + AT)x(t) + (BT)u(t)$$

We are able to remove that limit because in a discrete system, the time interval between samples is positive and non-negligible. By definition, a discrete system is only defined at certain time points, and not at all time points as the limit would have indicated. In a discrete system, we are interested only in the value of the system at discrete points. If those points are evenly spaced by every T seconds (the sampling time), then the samples of the system occur at $t = kT$, where k is an integer. Substituting kT for t into our equation above gives us:

$$x(kT + T) = (1 + AT)x(kT) + TBu(kT)$$

Or, using the square-bracket shorthand that we've developed earlier, we can write:

$$x[k+1] = (1 + AT)x[k] + TBu[k]$$

In this form, the state-space system can be implemented quite easily into a digital computer system using software, not complicated analog hardware.

We will write out the discrete-time state-space equations as:

$$x[n+1] = A_d x[n] + B_d u[n]$$
$$y[n] = C_d x[n] + D_d u[n]$$

Note on Notations

The variable T is a common variable in control systems, especially when talking about the beginning and end points of a continuous-time system, or when discussing the sampling time of a digital system. However, another common use of the letter T is to signify the transpose operation on a matrix. To alleviate this ambiguity, we will denote the transpose of a matrix with a *prime*:

$$A^T \rightarrow A'$$

Where A' is the transpose of matrix A.

The prime notation is also frequently used to denote the time-derivative. Most of the matrices that we will be talking about are time-invariant; there is no ambiguity because we will never take the time derivative of a time-invariant matrix. However, for a time-variant matrix we will use the following notations to distinguish between the time-derivative and the transpose:

$A(t)'$ the transpose.

$A'(t)$ the time-derivative.

Note that certain variables which are time-variant are not written with the *(t)* postscript, such as the variables x, y, and u. For these variables, the default behavior of the prime is the time-derivative, such as in the state equation. If the transpose needs to be taken of one of these vectors, the *(t)'* postfix will be added explicitly to correspond to our notation above.

For instances where we need to use the Hermitian transpose, we will use the notation:

$$A^H$$

This notation is common in other literature, and raises no obvious ambiguities here.

MATLAB Representation

State-space systems can be represented in MATLAB using the 4 system matrices, A, B, C, and D. We can create a system data structure using the ss function:

```
sys = ss(A, B, C, D);
```

Systems created in this way can be manipulated in the same way that the transfer function descriptions (described earlier) can be manipulated. To convert a transfer function to a state-space representation, we can use the tf2ss function:

```
[A, B, C, D] = tf2ss(num, den);
```

And to perform the opposite operation, we can use the ss2tf function:

```
[num, den] = ss2tf(A, B, C, D);
```

LINEAR SYSTEM SOLUTIONS

State Equation Solutions

The state equation is a first-order linear differential equation, or (more precisely) a system of linear differential equations. Because this is a first-order equation, we can use results from Ordinary Differential Equations to find a general solution to the equation in terms of the state-variable x. Once the state equation has been solved for x, that solution can be plugged into the output equation. The resulting equation will show the direct relationship between the system input and the system output, without the need to account explicitly for the internal state of the system.

Solving for x(t) With Zero Input

Looking again at the state equation:

$$x' = Ax(t) + Bu(t)$$

We can see that this equation is a first-order differential equation, except that the variables are vectors, and the coefficients are matrices. However, because of the rules of matrix calculus, these distinctions don't matter. We can ignore the input term (for now), and rewrite this equation in the following form:

$$\frac{dx(t)}{dt} = Ax(t)$$

And we can separate out the variables as such:

$$\frac{dx(t)}{x(t)} = Adt$$

Integrating both sides, and raising both sides to a power of e, we obtain the result:

$$x(t) = e^{At+C}$$

Where C is a constant. We can assign $D = e^C$ to make the equation easier, but we also know that D will then be the initial conditions of the system. This becomes obvious if we plug the value zero into the variable t. The final solution to this equation then is given as:

$$x(t) = e^{A(t-t_0)} x(t_0)$$

We call the matrix exponential e^{At} the state-transition matrix, and calculating it, while difficult at times, is crucial to analyzing and manipulating systems. We will talk more about calculating the matrix.

Solving for x(t) With Non-Zero Input

If, however, our input is non-zero (as is generally the case with any interesting system), our solution is a little bit more complicated. Notice that now that we have our input term in the equation, we will no longer be able to separate the variables and integrate both sides easily.

$$x'(t) = Ax(t) + Bu(t)$$

We subtract to get the $Ax(t)$ on the left side, and then we do something curious; we pre multiply both sides by the inverse state transition matrix:

$$e^{-At} x'(t) - e^{-At} Ax(t) = e^{-At} Bu(t)$$

The rationale for this last step may seem fuzzy at best, so we will illustrate the point with an example:

Example

Take the derivative of the following with respect to time:

$$e^{-At} x(t)$$

The product rule from differentiation reminds us that if we have two functions multiplied together:

$$f(t) g(t)$$

and we differentiate with respect to t, then the result is:

$$f(t) g'(t) + f'(t) g(t)$$

If we set our functions accordingly:

$$f(t) = e^{-At} \qquad f'(t) = -Ae^{-At}$$
$$g(t) = x(t) \qquad g'(t) = x'(t)$$

Then the output result is:

$$e^{-At} x'(t) - e^{-At} Ax(t)$$

If we look at this result, it is the same as from our equation above.

Using the result from our example, we can condense the left side of our equation into a derivative:

$$\frac{d(e^{-At}x(t))}{dt} = e^{-At}Bu(t)$$

Now we can integrate both sides, from the initial time (t_0) to the current time (t), using a dummy variable τ, we will get closer to our result. Finally, if we premultiply by e^{At}, we get our final result:

[General State Equation Solution]

$$x(t) = e^{A(t-t_0)}x(t_0) + \int_{t_0}^{t} e^{A(t-\tau)}Bu(\tau)d\tau$$

If we plug this solution into the output equation, we get:

[General Output Equation Solution]

$$y(t) = Ce^{A(t-t_0)}x(t_0) + C\int_{t_0}^{t} e^{A(t-\tau)}Bu(\tau)d\tau + Du(t)$$

This is the general Time-Invariant solution to the state space equations, with non-zero input. These equations are important results, and students who are interested in a further study of control systems would do well to memorize these equations.

State-Transition Matrix

The state transition matrix, e^{At}, is an important part of the general state-space solutions for the time-invariant cases listed above. Calculating this matrix exponential function is one of the very first things that should be done when analyzing a new system, and the results of that calculation will tell important information about the system in question.

The matrix exponential can be calculated directly by using a Taylor-Series expansion:

$$e^{At} = \sum_{n=0}^{\infty} \frac{(At)^n}{n!}$$

Also, we can attempt to diagonalize the matrix A into a diagonal matrix or a Jordan Canonical matrix. The exponential of a diagonal matrix is simply the diagonal elements individually raised to that exponential. The exponential of a Jordan canonical matrix is slightly more complicated, but there is a useful pattern that can be exploited to find the solution quickly. Interested readers should read the relevant passages in Engineering Analysis.

The state transition matrix, and matrix exponentials in general are very important tools in control engineering.

Diagonal Matrices

If a matrix is diagonal, the state transition matrix can be calculated by raising each diagonal entry of the matrix raised as a power of e.

Jordan Canonical Form

If the A matrix is in the Jordan Canonical form, then the matrix exponential can be generated quickly using the following formula:

$$e^{Jt} = e^{\lambda t} \begin{bmatrix} 1 & t & \frac{1}{2!}t^2 & \cdots & \frac{1}{n!}t^n \\ 0 & 1 & t & \cdots & \frac{1}{(n-1)!}t^{n-1} \\ \vdots & \vdots & \vdots & \ddots & \vdots \\ 0 & 0 & 0 & \cdots & 1 \end{bmatrix}$$

Where λ is the eigenvalue (the value on the diagonal) of the jordan-canonical matrix.

Inverse Laplace Method

We can calculate the state-transition matrix (or any matrix exponential function) by taking the following inverse Laplace transform:

$$e^{At} = \mathcal{L}^{-1}[(sI - A)^{-1}]$$

If A is a high-order matrix, this inverse can be difficult to solve.

If the A matrix is in the Jordan Canonical form, then the matrix exponential can be generated quickly using the following formula:

$$e^{Jt} = e^{\lambda t} \begin{bmatrix} 1 & t & \frac{1}{2!}t^2 & \cdots & \frac{1}{n!}t^n \\ 0 & 1 & t & \cdots & \frac{1}{(n-1)!}t^{n-1} \\ \vdots & \vdots & \vdots & \ddots & \vdots \\ 0 & 0 & 0 & \cdots & 1 \end{bmatrix}$$

Where λ is the eigenvalue (the value on the diagonal) of the jordan-canonical matrix.

Spectral Decomposition

If we know all the eigenvalues of A, we can create our transition matrix T, and our inverse transition matrix T^{-1} These matrices will be the matrices of the right and left eigenvectors, respectively. If we have both the left and the right eigenvectors, we can calculate the state-transition matrix as:

$$e^{At} = \sum_{i=1}^{n} e^{\lambda_i t} v_i w_i'$$

Note that w_i' is the transpose of the ith left-eigenvector, not the derivative of it.

Cayley-Hamilton Theorem

The Cayley-Hamilton Theorem can also be used to find a solution for a matrix exponential. For any eigenvalue of the system matrix A, λ, we can show that the two equations are equivalent:

$$e^{\lambda t} = a_0 + a_1 \lambda t + a_2 \lambda^2 t^2 + \cdots + a_{n-1} \lambda^{n-1} t^{n-1}$$

Once we solve for the coefficients of the equation, a, we can then plug those coefficients into the following equation:

$$e^{\lambda t} = a_0 + a_1 \lambda t + a_2 \lambda^2 t^2 + \cdots + a_{n-1} \lambda^{n-1} t^{n-1}$$

Example: Off-Diagonal Matrix

Given the following matrix A, find the state-transition matrix:

$$A = \begin{bmatrix} 0 & 1 \\ -1 & 0 \end{bmatrix}$$

We can find the eigenvalues of this matrix as $\lambda = i, -i$. If we plug these values into our eigenvector equation, we get:

$$\begin{vmatrix} i & -1 \\ 1 & i \end{vmatrix} v_1 = 0$$

$$\begin{vmatrix} -i & -1 \\ 1 & -i \end{vmatrix} v_2 = 0$$

And we can solve for our eigenvectors:

$$v_1 = \begin{bmatrix} 1 \\ i \end{bmatrix}$$

$$v_2 = \begin{bmatrix} 1 \\ -i \end{bmatrix}$$

With our eigenvectors, we can solve for our left-eigenvectors:

$$w_1 = \begin{bmatrix} 1 \\ -i \end{bmatrix}$$

$$w_2 = \begin{bmatrix} 1 \\ i \end{bmatrix}$$

Now, using spectral decomposition, we can construct the state-transition matrix:

$$e^{At} = e^{it} \begin{bmatrix} 1 \\ i \end{bmatrix} \begin{bmatrix} 1 & -i \end{bmatrix} + e^{-it} \begin{bmatrix} 1 \\ -i \end{bmatrix} \begin{bmatrix} 1 & i \end{bmatrix}$$

If we remember Euler's Identity, we can decompose the complex exponentials into sinusoids. Performing the vector multiplications, all the imaginary terms cancel out, and we are left with our result:

$$e^{At} = \begin{bmatrix} \cos t & \sin t \\ -\sin t & \cos t \end{bmatrix}$$

The reader is encouraged to perform the multiplications, and attempt to derive this result.

Example: MATLAB Calculation

Using the symbolic toolbox in MATLAB, we can write MATLAB code to automatically generate the state-transition matrix for a given input matrix A. Here is an example of MATLAB code that can perform this task:

```
function [phi] = statetrans(A)
    t = sym('t');
    phi = expm(A * t);
end
```

Use this MATLAB function to find the state-transition matrix for the following matrices (warning, calculation may take some time):

1. $A_1 = \begin{bmatrix} 2 & 0 \\ 0 & 2 \end{bmatrix}$

2. $A_2 = \begin{bmatrix} 0 & 1 \\ -1 & 0 \end{bmatrix}$

3. $A_3 = \begin{bmatrix} 2 & 1 \\ 0 & 2 \end{bmatrix}$

Matrix 1 is a diagonal matrix, Matrix 2 has complex eigenvalues, and Matrix 3 is Jordan canonical form. These three matrices should be representative of some of the common forms of system matrices. The following code snippets are the input commands into MATLAB to produce these matrices, and the output results:

Matrix A1

```
>> A1 = [2 0 ; 0 2];
>> statetrans(A1)
ans =


[ exp(2*t),          0]
[          0, exp(2*t)]
```

Matrix A2

```
>> A2 = [0 1 ; -1 0];
>> statetrans(A1)

ans =

[  cos(t),    sin(t)]
[ -sin(t),    cos(t)]
```

Matrix A3

```
>> A1 = [2 1 ; 0 2];
>> statetrans(A1)

ans =

[   exp(2*t), t*exp(2*t)]
[          0,   exp(2*t)]
```

Example: Multiple Methods in MATLAB

There are multiple methods in MATLAB to compute the state transtion matrix, from a scalar (time-invariant) matrix A. The following methods are all going to rely on the *Symbolic Toolbox* to perform the equation manipulations. At the end of each code snippet, the variable eAt contains the state-transition matrix of matrix A.

Direct Method

```
t = sym('t');
eAt = expm(A * t);
```

Laplace Transform Method

```
s = sym('s');
[n,n] = size(A);
in = inv(s*eye(n) - A);
eAt = ilaplace(in);
```

Spectral Decomposition

```
t = sym('t');
[n,n] = size(A);
[V, e] = eig(A);
W = inv(V);
sum = [0 0;0 0];
for I = 1:n
    sum = sum + expm(e(I,I)*t)*V(:,I)*W(I,:);
end;
eAt = sum;
```

All three of these methods should produce the same answers. The student is encouraged to verify this.

TIME VARIANT SYSTEM SOLUTIONS

General Time Variant Solution

The state-space equations can be solved for time-variant systems, but the solution is significantly more complicated than the time-invariant case. Our time-variant state equation is given as follows:

$$x'(t) = A(t)\, x(t) + B(t)u(t)$$

We can say that the general solution to time-variant state-equation is defined as:

[Time-Variant General Solution]

$$x(t) = \phi(t, t_0)\, x(t_0) + \int_{t_0}^{t} \phi(t, \tau)\, B(\tau)\, u(\tau)\, d\tau$$

Matrix Dimensions:

$$A: p \times p$$
$$B: p \times q$$
$$C: r \times p$$
$$D: r \times q$$

The function φ is called the state-transition matrix, because it (like the matrix exponential from the time-invariant case) controls the change for states in the state equation. However, unlike the time-invariant case, we cannot define this as a simple exponential. In fact, φ can't be defined in general, because it will actually be a different function for every system. However, the state-transition matrix does follow some basic properties that we can use to determine the state-transition matrix.

In a time-variant system, the general solution is obtained when the state-transition matrix is determined. For that reason, the first thing (and the most important thing) that we need to do here is find that matrix.

State Transition Matrix

The state transition matrix φ is not completely unknown, it must always satisfy the following relationships:

$$\frac{\partial \phi(t, t_0)}{\partial t} = A(t)\, \phi(t, t_0)$$

$$\phi(\tau, \tau) = I$$

And φ also must have the following properties:

1. $\phi(t_2, t_1)\phi(t_1, t_0) = \phi(t_2, t_0)$

2. $\phi^{-1}(t, \tau) = \phi(\tau, t)$

3. $\phi^{-1}(t, \tau)\, \phi(\tau, t) = I$

4. $\dfrac{d\phi(t_0, t_0)}{dt} = A(t)$

If the system is time-invariant, we can define φ as:

$$\phi(t, t_0) = e^A (t - t_0)$$

The reader can verify that this solution for a time-invariant system satisfies all the properties listed above. However, in the time-variant case, there are many different functions that may satisfy these requirements, and the solution is dependant on the structure of the system. The state-transition matrix must be determined before analysis on the time-varying solution can continue.

Time-Variant, Zero Input

As the most basic case, we will consider the case of a system with zero input. If the system has no input, then the state equation is given as:

$$x'(t) = A(t)x(t)$$

And we are interested in the response of this system in the time interval $T = (a, b)$. The first thing we want to do in this case is find a fundamental matrix of the above equation. The fundamental matrix is related

Fundamental Matrix

Here, x is an $n \times 1$ vector, and A is an $n \times n$ matrix.

Given the equation:

$$x'(t) = A(t)x(t)$$

The solutions to this equation form an n-dimensional vector space in the interval T = (a, b). Any set of n linearly-independent solutions $\{x_1, x_2,..., x_n\}$ to the equation above is called a fundamental set of solutions.

Readers who have a background in Linear Algebra may recognize that the fundamental set is a basis set for the solution space. Any basis set that spans the entire solution space is a valid fundamental set.

A fundamental matrix FM is formed by creating a matrix out of the n fundamental vectors. We will denote the fundamental matrix with a script capital X:

$$\chi = \begin{bmatrix} x_1 & x_2 & \cdots & x_n \end{bmatrix}$$

The fundamental matrix will satisfy the state equation:

$$\chi'(t) = A(t)\, \chi(t)$$

Also, *any matrix that solves this equation can be a fundamental matrix if* and only if the determinant of the matrix is non-zero for all time t in the interval T. The determinant must be non-zero, because we are going to use the inverse of the fundamental matrix to solve for the state-transition matrix.

State Transition Matrix

Once we have the fundamental matrix of a system, we can use it to find the state transition matrix of the system:

$$\phi(t, t_0) = \chi(t)\, \chi^{-1}(t_0)$$

The inverse of the fundamental matrix exists, because we specify in the definition above that it must have a non-zero determinant, and therefore must be non-singular. The reader should note that this is only one possible method for determining the state transition matrix.

Example: 2-Dimensional System

Given the following fundamental matrix, Find the state-transition matrix.

$$\chi(t) = \begin{bmatrix} e^{-1} & \frac{1}{2}e^{t} \\ 0 & e^{-t} \end{bmatrix}$$

the first task is to find the inverse of the fundamental matrix. Because the fundamental matrix is a 2 × 2 matrix, the inverse can be given easily through a common formula:

$$\chi^{-1}(t) = \begin{bmatrix} e^{-1} & -\dfrac{1}{2}e^{t} \\ 0 & e^{-t} \end{bmatrix} = \begin{bmatrix} e^{t} & -\dfrac{1}{2}e^{3t} \\ 0 & e^{t} \end{bmatrix}$$

The state-transition matrix is given by:

$$\phi(t, t_0) = \chi(t)\,\chi^{-1}(t_0) = \begin{bmatrix} e^{-t} & -\dfrac{1}{2}e^{t} \\ 0 & e^{-t} \end{bmatrix} = \begin{bmatrix} e^{t_0} & \dfrac{1}{2}e^{3t_0} \\ 0 & e^{t_0} \end{bmatrix}$$

$$\phi(t, t_0) = \begin{bmatrix} e^{t+t_0} & \dfrac{1}{2}(e^{t+t_0} - e^{-t+3t_0}) \\ 0 & e^{-t+t_0} \end{bmatrix}$$

Other Methods

There are other methods for finding the state transition matrix besides having to find the fundamental matrix.

Method 1

If A(t) is triangular (upper or lower triangular), the state transition matrix can be determined by sequentially integrating the individual rows of the state equation.

Method 2

If for every τ and t, the state matrix commutes as follows:

$$A(t)\left[\int_\tau^t A(\zeta)\,d\zeta\right] = \left[\int_\tau^t A(\zeta)\,d\zeta\right]A(t)$$

Then the state-transition matrix can be given as:

$$\phi(t, \tau) = e^{\int_\tau^t A(\zeta)\,d\zeta}$$

The state transition matrix will commute as described above if any of the following conditions are true:

1. A is a constant matrix (time-invariant)
2. A is a diagonal matrix

3. If $A = \overline{A}f(t)$, where $\overline{A}$ is a constant matrix, and f(t) is a single-valued function (not a matrix).

If none of the above conditions are true, then you must use method 3.

Method 3

If $A(t)$ can be decomposed as the following sum:

$$A(t) = \sum_{i=1}^{n} M_i f_i(t)$$

Where M_i is a constant matrix such that $M_i M_j = M_j M_i$, and f_i is a single-valued function. If $A(t)$ can be decomposed in this way, then the state-transition matrix can be given as:

$$\phi(t, \tau) = \prod_{i=1}^{n} e^{M_i \int_{\tau}^{t} f_i(\theta) d\theta}$$

It will be left as an exercise for the reader to prove that if $A(t)$ is time-invariant, that the equation in method 2 above will reduce to the state-transition matrix $e^{A(t-\tau)}$.

Example: Using Method 3

Use method 3, above, to compute the state-transition matrix for the system if the system matrix A is given by:

$$A = \begin{bmatrix} t & 1 \\ -1 & t \end{bmatrix}$$

We can decompose this matrix as follows:

$$A = \begin{bmatrix} 1 & 0 \\ 0 & 1 \end{bmatrix} t + \begin{bmatrix} 0 & 1 \\ -1 & 0 \end{bmatrix}$$

Where $f_1(t) = t$, and $f_2(t) = 1$. Using the formula described above gives us:

$$\phi(t, \tau) = e^{M_1 \int_{\tau}^{t} \theta d\theta} e^{M_2 \int_{\tau}^{t} d\theta}$$

Solving the two integrations gives us:

$$\phi(t, \tau) = e^{\frac{1}{2}\begin{bmatrix} (t^2 - \tau^2) & 0 \\ 0 & (t^2 - \tau^2) \end{bmatrix}} e^{\begin{bmatrix} 0 & t-\tau \\ -t-\tau & 0 \end{bmatrix}}$$

The first term is a diagonal matrix, and the solution to that matrix function is all the individual elements of the matrix raised as an exponent of e. The second term can be decomposed as:

$$e^{\begin{bmatrix} 0 & t-\tau \\ -t+\tau & 0 \end{bmatrix}} = e^{\begin{bmatrix} 0 & 1 \\ -1 & 0 \end{bmatrix}(t-\tau)} = \begin{bmatrix} \cos(t-\tau) & \sin(t-\tau) \\ -\sin(t-\tau) & \cos(t-\tau) \end{bmatrix}$$

The final solution is given as:

$$\phi(t,\tau) = \begin{bmatrix} e^{\frac{1}{2}(t^2-\tau^2)} & 0 \\ 0 & e^{\frac{1}{2}(t^2-\tau^2)} \end{bmatrix} \begin{bmatrix} \cos(t-\tau) & \sin(t-\tau) \\ -\sin(t-\tau) & \cos(t-\tau) \end{bmatrix}$$

$$= \begin{bmatrix} e^{\frac{1}{2}(t^2-\tau^2)}\cos(t-\tau) & e^{\frac{1}{2}(t^2-\tau^2)}\sin(t-\tau) \\ -e^{\frac{1}{2}(t^2-\tau^2)}\sin(t-\tau) & e^{\frac{1}{2}(t^2-\tau^2)}\cos(t-\tau) \end{bmatrix}$$

Time-Variant, Non-zero Input

If the input to the system is not zero, it turns out that all the analysis that we performed above still holds. We can still construct the fundamental matrix, and we can still represent the system solution in terms of the state transition matrix φ.

We can show that the general solution to the state-space equations is actually the solution:

$$x(t) = \phi(t,t_0)\,x(t_0) + \int_{t_0}^{t} \phi(t,\tau)\,B(\tau)\,u(\tau)\,d\tau$$

EIGENVALUES AND EIGENVECTORS

Eigenvalues and Eigenvectors

The eigenvalues and eigenvectors of the system matrix play a key role in determining the response of the system. It is important to note that only square matrices have eigenvalues and eigenvectors associated with them. Non-square matrices cannot be analyzed using the methods below.

The word "eigen" is from the German for "characteristic", and so this chapter could also be called "Characteristic values and characteristic vectors". The terms "Eigenvalues" and "Eigenvectors" are most commonly used. Eigenvalues and Eigenvectors have a number of properties that make them valuable tools in analysis, and they also have a number of valuable relationships with the matrix from which they are derived. Computing the eigenvalues and the eigenvectors of the system matrix is one of the most important things that should be done when beginning to analyze a system matrix, second only to calculating the matrix exponential of the system matrix.

The eigenvalues and eigenvectors of the system determine the relationship between the individual system state variables (the members of the x vector), the response of the system to inputs, and the stability of the system. Also, the eigenvalues and eigenvectors can be used to calculate the matrix exponential of the system matrix (through spectral decomposition).

Characteristic Equation

The characteristic equation of the system matrix A is given as:

[Matrix Characteristic Equation]

$$Av = \lambda v$$

Where λ are scalar values called the eigenvalues, and v are the corresponding eigenvectors. To solve for the eigenvalues of a matrix, we can take the following determinant:

$$|A - \lambda I| = 0$$

To solve for the eigenvectors, we can then add an additional term, and solve for v:

$$(A - \lambda I)v = 0$$

Another value worth finding are the left eigenvectors of a system, defined as w in the modified characteristic equation:

[Left-Eigenvector Equation]

$$wA = \lambda w$$

Diagonalization

If the matrix A has a complete set of distinct eigenvalues, the matrix can be diagonalized. A diagonal matrix is a matrix that only has entries on the diagonal, and all the rest of the entries in the matrix are zero. We can define a transformation matrix, T, that satisfies the diagonalization transformation:

$$A = TDT^{-1}$$

Which in turn will satisfy the relationship:

$$e^{At} = Te^{Dt}T^{-1}$$

The right-hand side of the equation may look more complicated, but because D is a diagonal matrix here (not to be confused with the feed-forward matrix from the output equation), the calculations are much easier.

We can define the transition matrix, and the inverse transition matrix in terms of the eigenvectors and the left eigenvectors:

$$T = \begin{bmatrix} v_1 & v_2 & v_3 & \cdots & v_n \end{bmatrix}$$

$$T^{-1} = \begin{bmatrix} w'_1 \\ w'_2 \\ w'_3 \\ \vdots \\ w'_n \end{bmatrix}$$

Exponential Matrix Decomposition

A matrix exponential can be decomposed into a sum of the eigenvectors, eigenvalues, and left eigenvectors, as follows:

$$e^{At} = \sum_{i=1}^{n} e^{\lambda_i t} v_i w_i'$$

Notice that this equation only holds in this form if the matrix A has a complete set of n distinct eigenvalues. Since w'_i is a row vector, and $x(0)$ is a column vector of the initial system states, we can combine those two into a scalar coefficient α:

$$e^{At} x(t_0) = \sum_{i=1}^{n} \alpha_i e^{\lambda_i t} v_i$$

Since the state transition matrix determines how the system responds to an input, we can see that the system eigenvalues and eigenvectors are a key part of the system response. Let us plug this decomposition into the general solution to the state equation:

[State Equation Spectral Decomposition]

$$x(t) = \sum_{i=1}^{n} \alpha_i e^{\lambda_i t} v_i + \sum_{i=1}^{n} \int_0^t e^{\lambda_i (t-\tau)} v_i w_i' Bu(\tau) d\tau$$

State Relationship

As we can see from the above equation, the individual elements of the state vector $x(t)$ cannot take arbitrary values, but they are instead related by weighted sums of multiples of the systems right-eigenvectors.

Decoupling

For people who are familiar with linear algebra, the left-eigenvector of the matrix A must be in the *null space* of the matrix B to decouple the system.

If a system can be designed such that the following relationship holds true:

$$w_i' B = 0$$

then the system response from that particular eigenvalue will not be affected by the system input u, and we say that the system has been decoupled. Such a thing is difficult to do in practice.

Condition Number

With every matrix there is associated a particular number called the condition number of that matrix. The condition number tells a number of things about a matrix, and it is worth calculating. The condition number, k, is defined as:

$$k = \frac{\|w_i\| \|v_i\|}{|w_i' v_i|}$$

Systems with smaller condition numbers are better, for a number of reasons:

1. Large condition numbers lead to a large transient response of the system
2. Large condition numbers make the system eigenvalues more sensitive to changes in the system.

Stability

Notice that if the eigenvalues of the system matrix A are *positive*, or (if they are complex) that they have positive real parts, that the system state (and therefore the system output, scaled by the C matrix) will approach infinity as time *t* approaches infinity. In essence, if the eigenvalues are positive, the system will not satisfy the condition of BIBO stability, and will therefore become *unstable*.

Another factor that is worth mentioning is that a manufactured system *never exactly matches the system model*, and there will always been inaccuracies in the specifications of the component parts used, *within a certain tolerance*. As such, the system matrix will be slightly different from the mathematical model of the system (although good systems will not be severely different), and therefore the eigenvalues and eigenvectors of the system will not be the same values as those derived from the model. These facts give rise to several results:

1. Systems with high *condition numbers* may have eigenvalues that differ by a large amount from those derived from the mathematical model. This means that the system response of the physical system may be very different from the intended response of the model.
2. Systems with high condition numbers may become *unstable* simply as a result of inaccuracies in the component parts used in the manufacturing process.

For those reasons, the system eigenvalues and the condition number of the system matrix are highly important variables to consider when analyzing and designing a system.

Non-Unique Eigenvalues

The decomposition above only works if the matrix A has a full set of n distinct eigenvalues (and corresponding eigenvectors). If A does not have n distinct eigenvectors, then a set of generalized eigenvectors need to be determined. The generalized eigenvectors will produce a similar matrix that is in Jordan canonical form, not the diagonal form we were using earlier.

Generalized Eigenvectors

Generalized eigenvectors can be generated using the following equation:

$$(A - \lambda I)\, v_{n+1} = v_n$$

if d is the number of times that a given eigenvalue is repeated, and p is the number of unique eigenvectors derived from those eigenvalues, then there will be

$$q = d - p$$

generalized eigenvectors. Generalized eigenvectors are developed by plugging in the regular eigenvectors into the equation above (v_n). Some regular eigenvectors might not produce any non-trivial generalized eigenvectors. Generalized eigenvectors may also be plugged into the equation above to produce additional generalized eigenvectors. It is important to note that the generalized eigenvectors form an ordered series, and they must be kept in order during analysis or the results will not be correct.

Example: One Repeated Set

We have a 5 × 5 matrix A with eigenvalues λ = 1, 1, 1, 2, 2. For λ = 1, there is 1 distinct eigenvector a. For λ = 2 there is 1 distinct eigenvector b. From a, we generate the generalized eigenvector c, and from c we can generate vector d. From the eigevector b, we generate the generalized eigevector e. In order our eigenvectors are listed as:

$$[a\ c\ d\ b\ e]$$

Notice how c and d are listed in order after the eigenvector that they are generated from, a. Also, we could reorder this as:

$$[b\ e\ a\ c\ d]$$

because the generalized eigenvectors are listed in order after the regular eigenvector that they are generated from. Regular eigenvectors can be listed in any order.

Example: Two Repeated Sets

We have a 4 × 4 matrix A with eigenvalues λ = 1, 1, 1, 2. For λ = 1 we have two eigevectors, a and b. For λ = 2 we have an eigenvector c.

We need to generate a fourth eigenvector, d. The only eigenvalue that needs another eigenvector is λ = 1, however there are already two eigevectors associated with that eigenvalue, and only one of them will generate a non-trivial generalized eigenvector. To figure out which one works, we need to plug both vectors into the generating equation:

$$(A - \lambda I)\big|_{\lambda=1} d = a$$

$$(A - \lambda I)\big|_{\lambda=1} d = b$$

If a generates the correct vector d, we will order our eigenvectors as:

$$[a\ d\ b\ c]$$

but if b generates the correct vector, we can order it as:

$$[a\ b\ d\ c]$$

Jordan Canonical Form

If a matrix has a complete set of distinct eigenvectors, the transition matrix T can be defined as the matrix of those eigenvectors, and the resultant transformed matrix will be a diagonal matrix. However, if the eigenvectors are not unique, and there are a number of generalized eigenvectors associated with the matrix, the transition matrix T will consist of the ordered set of the regular eigenvectors and generalized eigenvectors. The regular eigenvectors that did not produce any generalized eigenvectors (if any) should be first in the order, followed by the eigenvectors that did produce generalized eigenvectors, and the generalized eigenvectors that they produced (in appropriate sequence).

Once the T matrix has been produced, the matrix can be transformed by it and it's inverse:

$$A = T^{-1} JT$$

The J matrix will be a Jordan block matrix. The format of the Jordan block matrix will be as follows:

$$J = \begin{bmatrix} D & 0 & \cdots & 0 \\ 0 & J_1 & \cdots & 0 \\ \vdots & \vdots & \ddots & \vdots \\ 0 & 0 & \cdots & J_n \end{bmatrix}$$

Where D is the diagonal block produced by the regular eigenvectors that are not associated with generalized eigenvectors (if any). The J_n blocks are standard Jordan blocks with a size corresponding to the number of eigenvectors/generalized eigenvectors in each sequence. In each J_n block, the eigenvalue associated with the regular eigenvector of the sequence is on the main diagonal, and there are 1's in the sub-diagonal.

System Response

Equivalence Transformations

If we have a non-singular $n \times n$ matrix P, we can define a transformed vector "x bar" as:

$$\bar{x} = Px$$

We can transform the entire state-space equation set as follows:

$$\bar{x}'(t) = \bar{A}\bar{x}(t) + \bar{B}u(t)$$
$$\bar{y}(t) = \bar{C}\bar{x}(t) + \bar{D}u(t)$$

Where:

$$\bar{A} = PAP^{-1}$$

$$\overline{B} = PB$$
$$\overline{C} = CP^{-1}$$
$$\overline{D} = D$$

We call the matrix P the equivalence transformation between the two sets of equations.

It is important to note that the eigenvalues of the matrix A (which are of primary importance to the system) do not change under the equivalence transformation. The eigenvectors of A, and the eigenvectors of $\overline{A}$ are related by the matrix P.

Lyapunov Transformations

The transformation matrix P is called a Lyapunov Transformation if the following conditions hold:

- P(t) is nonsingular.
- P(t) and P'(t) are continuous
- P(t) and the inverse transformation matrix P-1(t) are finite for all t.

If a system is time-variant, it can frequently be useful to use a Lyapunov transformation to convert the system to an equivalent system with a constant A matrix. This is not always possible in general, however it is possible if the $A(t)$ matrix is periodic.

System Diagonalization

If the A matrix is time-invariant, we can construct the matrix V from the eigenvectors of A. The V matrix can be used to transform the A matrix to a diagonal matrix. Our new system becomes:

$$Vx'(t) = V\,AV^{-1}\,Vx(t) + VBu(t)$$
$$y(t) = CV^{-1}\,Vs(t) + Du(t)$$

Since our system matrix is now diagonal (or Jordan canonical), the calculation of the state-transition matrix is simplified:

$$e^{V\,AV^{-1}} = \Lambda$$

Where Λ is a diagonal matrix.

MATLAB Transformations

The MATLAB function ss2ss can be used to apply an equivalence transformation to a system. If we have a set of matrices A, B, C and D, we can create equivalent matrices as such:

```
[Ap, Bp, Cp, Dp] = ss2ss(A, B, C, D, p);
```

Where p is the equivalence transformation matrix.

STANDARD FORMS

Companion Form

A companion form contains the coefficients of a corresponding characteristic polynomial along one of its far rows or columns. For example, one companion form matrix is:

$$
\begin{bmatrix}
0 & 0 & 0 & \cdots & 0 & -a_0 \\
1 & 0 & 0 & \cdots & 0 & -a_1 \\
0 & 1 & 0 & \cdots & 0 & -a_2 \\
0 & 0 & 1 & \cdots & 0 & -a_3 \\
\vdots & \vdots & \vdots & \ddots & \vdots & \vdots \\
0 & 0 & 0 & \cdots & 1 & -a_{n-1}
\end{bmatrix}
$$

and another is:

$$
\begin{bmatrix}
-a_{n-1} & -a_{n-2} & -a_{n-3} & \cdots & -a_1 & -a_0 \\
1 & 0 & 0 & \cdots & 0 & 0 \\
0 & 1 & 0 & \cdots & 0 & 0 \\
0 & 0 & 1 & \cdots & 0 & 0 \\
\vdots & \vdots & \vdots & \ddots & \vdots & \vdots \\
0 & 0 & 0 & \cdots & 1 & 0
\end{bmatrix}
$$

There are two companion forms that are convenient to use in control theory, namely the observable canonical form and the controllable canonical form. These two forms are roughly transposes of each other (just as observability and controllability are dual ideas). When placed in one of these forms, the design of controllers or observers is simplified because the structure of the system is made apparent (and is easily modified with the desired control).

Observable Canonical Form

Observable-Canonical Form is useful in a number of cases, especially for designing observers.

The observable-canonical form is as follows:

$$
A = \begin{bmatrix}
-a_1 & 1 & 0 & \cdots & 0 \\
-a_2 & 0 & 1 & \cdots & 0 \\
\vdots & \vdots & \vdots & \ddots & \vdots \\
-a_{n-1} & 0 & 0 & \cdots & 1 \\
-a_n & 0 & 0 & \cdots & 0
\end{bmatrix}
$$

$$B = \begin{bmatrix} b_1 \\ b_2 \\ \vdots \\ b_n \end{bmatrix}$$

$$C = \begin{bmatrix} 1 & 0 & \cdots & 0 \end{bmatrix}$$

Controllable Canonical Form

Controllable-Canonical Form is useful in a number of cases, especially for designing controllers when the full state of the system is known.

The controllable-canonical form is as follows:

$$A = \begin{bmatrix} -a_1 & -a_2 & -a_3 & \cdots & -a_{n-1} & -a_n \\ 1 & 0 & 0 & \cdots & 0 & 0 \\ 0 & 1 & 0 & \cdots & 0 & 0 \\ 0 & 0 & 1 & \cdots & 0 & 0 \\ \vdots & \vdots & \vdots & \ddots & \vdots & \vdots \\ 0 & 0 & 0 & \cdots & 1 & 0 \end{bmatrix}$$

$$B = \begin{bmatrix} 1 \\ 0 \\ \vdots \\ 0 \end{bmatrix}$$

$$C = \begin{bmatrix} b_1 & b_2 & b_3 & \cdots & b_n \end{bmatrix}$$

$$D = \begin{bmatrix} b_0 \end{bmatrix}$$

If we have two spaces, space v which is the original space of the system (A, B, C, and D), then we can transform our system into the w space which is in controllable-canonical form (A_w, B_w, C_w, D_w) using a transformation matrix T_w. We define this transformation matrix as:

$$T = \zeta_v \zeta_w^{-1}$$

Where ζ is the controlability matrix.

Notice that we know beforehand A_w and B_w, since we know both the form of the matrices and the coefficients of the equation (e.g. a linear ODE with constant coefficients or a transfer function).

If we know these two matrices, then we can form ζ_w. We can then use this matrix to create our transformation matrix.

The controllable canonical form later when we discuss state-feedback and closed-loop systems.

Phase Variable Form

The Phase Variable Form is obtained simply by renumbering the phase variables in the opposite order of the controllable canonical form. Thus:

$$A = \begin{bmatrix} 0 & 0 & 0 & \cdots & 1 & 0 \\ \vdots & \vdots & \vdots & \ddots & \vdots & \vdots \\ 0 & 0 & 1 & \cdots & 0 & 0 \\ 0 & 1 & 0 & \cdots & 0 & 0 \\ 1 & 0 & 0 & \cdots & 0 & 0 \\ -a_1 & -a_2 & -a_3 & \cdots & -a_{n-1} & -a_n \end{bmatrix}$$

$$B = \begin{bmatrix} 0 \\ 0 \\ \vdots \\ 1 \end{bmatrix}$$

$$C = \begin{bmatrix} b_n & b_{n-1} & \cdots & b_2 & b_1 \end{bmatrix}$$

$$D = \begin{bmatrix} b_0 \end{bmatrix}$$

Modal Form

In this form, the state matrix is a diagonal matrix of its (non-repeated) eigenvalues. The control has a unitary influence on each eigenspace, and the output is a linear combination of the contributions from the eigenspaces (where the weights are the complex residuals at each pole).

$$A = \begin{bmatrix} -p_1 & 0 & 0 & \cdots & 0 & 0 \\ 0 & -p_2 & 0 & \cdots & 0 & 0 \\ 0 & 0 & -p_3 & \cdots & 0 & 0 \\ \vdots & \vdots & \vdots & \ddots & \vdots & \vdots \\ 0 & 0 & 0 & \cdots & 0 & -p_n \end{bmatrix}$$

$$B = \begin{bmatrix} 1 \\ 1 \\ \vdots \\ 1 \end{bmatrix}$$

$$C = \begin{bmatrix} c_1 & c_2 & \cdots & c_n \end{bmatrix}$$

Jordan Form

This "almost diagonal" form handles the case where eigenvalues are repeated. The repeated eigenvalues represent a multi-dimensional eigenspace, and so the control only enters the eigenspace once and its integrated through the other states of that small subsystem.

$$A = \begin{bmatrix} -p_1 & 1 & 0 & 0 & 0 & \cdots & 0 & 0 \\ 0 & -p_1 & 1 & 0 & 0 & \cdots & 0 & 0 \\ 0 & 0 & -p_1 & 0 & 0 & \cdots & 0 & 0 \\ 0 & 0 & 0 & -p_4 & 0 & \cdots & 0 & 0 \\ \vdots & \vdots & \vdots & \vdots & \vdots & \ddots & \vdots & \vdots \\ 0 & 0 & 0 & 0 & 0 & \cdots & 0 & -p_n \end{bmatrix}$$

$$B = \begin{bmatrix} 0 \\ 0 \\ 1 \\ 1 \\ \vdots \\ 1 \end{bmatrix}$$

$$C = \begin{bmatrix} c_1 & c_2 & \cdots & c_n \end{bmatrix}$$

Computing Standard Forms in MATLAB

MATLAB contains a function for automatically transforming a state space equation into a companion (e.g., controllable or observable canonical form) form.

```
[Ap, Bp, Cp, Dp, P] = canon(A, B, C, D, 'companion');
```

This operation can be performed using this MATLAB command:

compan

Moving from one companion form to the other usually involves elementary operations on matrices and vectors (e.g., transposes or interchanging rows). Given a vector with the coefficients of a characteristic polynomial, MATLAB can compute the corresponding companion form.

```
compan(P)
```

Given another vector with the coefficients of a transfer function's numerator polynomial, the `canon` command can do the same.

```
[Ap, Bp, Cp, Dp, P] = canon(tf(Pnum,Pden), 'companion');
```

To transform a state space equation into a modal (e.g., diagonal) form, the same command can be used.

```
[Ap, Bp, Cp, Dp, P] = canon(A, B, C, D, 'modal');
```

This operation can be performed using thisMATLAB command:

jordan

However, MATLAB also includes a command to compute the Jordan form of a matrix, which is a modified modal form suited for matrices that have repeated eigenvalues.

```
jordan(A)
```

MIMO SYSTEMS

Multi-Input, Multi-Output

Systems with more than one input and/or more than one output are known as Multi-Input Multi-Output systems, or they are frequently known by the abbreviation MIMO. This is in contrast to systems that have only a single input and a single output (SISO).

State-Space Representation

MIMO systems that are lumped and linear can be described easily with state-space equations. To represent multiple inputs we expand the input $u(t)$ into a vector $U(t)$ with the desired number of inputs. Likewise, to represent a system with multiple outputs, we expand $y(t)$ into $Y(t)$, which is a vector of all the outputs. For this method to work, the outputs must be linearly dependant on the input vector and the state vector.

$$X'(t) = AX(t) + BU(t)$$
$$Y(t) = CX(t) + DU(t)$$

Example: Two Inputs and Two Outputs

Let's say that we have two outputs, y_1 and y_2, and two inputs, u_1 and u_2. These are related in our system through the following system of differential equations:

$$y_1'' + a_1 y_1' + a_0 (y_1 + y_2) = u_1(t)$$
$$y_2' + a_2 (y_2 + y_1) = u_2(t)$$

now, we can assign our state variables as such, and produce our first-order differential equations:

$$x_1 = y_1$$
$$x_4 = y_2$$
$$x_1' = y_1' = x_2$$
$$x_2' = -a_1 x_2 - a_0 (x_1 + x_4) + u_1(t)$$
$$x_4' = -a_2 (x_4 - x_1) + u_2(t)$$

And finally we can assemble our state space equations:

$$x' = \begin{bmatrix} 0 & 1 & 0 & 0 \\ -a_0 & -a_1 & 0 & -a_0 \\ 0 & 0 & 0 & 1 \\ a_2 & 0 & 0 & -a_2 \end{bmatrix} x + \begin{bmatrix} 0 & 0 \\ 1 & 0 \\ 0 & 0 \\ 0 & 1 \end{bmatrix} \begin{bmatrix} u_1 \\ u_2 \end{bmatrix}$$

$$\begin{bmatrix} y_1 \\ y_2 \end{bmatrix} = \begin{bmatrix} 1 & 0 & 0 & 0 \\ 0 & 0 & 0 & 1 \end{bmatrix} x(t)$$

Transfer Function Matrix

If the system is LTI and Lumped, we can take the Laplace Transform of the state-space equations, as follows:

$$\mathcal{L}[x'(t)] = \mathcal{L}[AX(t)] + \mathcal{L}[BU(t)]$$
$$\mathcal{L}[Y(t)] = \mathcal{L}[CX(t)] + \mathcal{L}[DU(t)]$$

Which gives us the result:

$$sX(s) - X(0) = AX(s) + BU(s)$$
$$Y(s) = CX(s) + DU(s)$$

Where $X(0)$ is the initial conditions of the system state vector in the time domain. If the system is relaxed, we can ignore this term, but for completeness we will continue the derivation with it.

We can separate out the variables in the state equation as follows:

$$sX(s) - AX(s) = X(0) + BU(s)$$

Then factor out an $X(s)$:

$$X(s)[sI - A] = X(0) + BU(s)$$

And then we can multiply both sides by the inverse of [sI - A] to give us our state equation:

$$X(s) = [sI - A]^{-1} X(0) + [sI - A]^{-1} BU(s)$$

Now, if we plug in this value for $X(s)$ into our output equation, above, we get a more complicated equation:

$$Y(s) = C([sI - A]^{-1} X(0) + [sI - A]^{-1} BU(s)) + DU(s)$$

And we can distribute the matrix C to give us our answer:

$$Y(s) = C[sI - A]^{-1} X(0) + C[sI - A]^{-1} BU(s) + DU(s)$$

Now, if the system is relaxed, and therefore $X(0)$ is 0, the first term of this equation becomes 0. In this case, we can factor out a $U(s)$ from the remaining two terms:

$$Y(s) = (C[sI - A]^{-1} B + D) U(s)$$

We can make the following substitution to obtain the Transfer Function Matrix, or more simply, the Transfer Matrix, $H(s)$:

$$C[sI - A]^{-1} B + D = H(s)$$

And rewrite our output equation in terms of the transfer matrix as follows:

$$Y(s) = H(s) U(s)$$

If $Y(s)$ and $X(s)$ are 1×1 vectors (a SISO system), then we have our external description:

$$Y(s) = H(s) X(s)$$

Now, since $X(s) = X(s)$, and $Y(s) = Y(s)$, then $H(s)$ must be equal to $H(s)$. These are simply two different ways to describe the same exact equation, the same exact system.

Dimensions

If our system has q inputs, and r outputs, our transfer function matrix will be an $r \times q$ matrix.

Relation to Transfer Function

For SISO systems, the Transfer Function matrix will reduce to the transfer function as would be obtained by taking the Laplace transform of the system response equation.

For MIMO systems, with n inputs and m outputs, the transfer function matrix will contain $n \times m$ transfer functions, where each entry is the transfer function relationship between each individual input, and each individual output.

Through this derivation of the transfer function matrix, we have shown the equivalency between the Laplace methods and the State-Space method for representing systems. Also, we have shown how the Laplace method can be generalized to account for MIMO systems. Through the rest of this book, we will use the Laplace and State Space methods interchangeably, opting to use one or the other where appropriate.

Zero-State and Zero-Input

If we have our complete system response equation from above:

$$Y(s) = C[sI - A]^{-1} x(0) + (C[sI - A]^{-1} B + D) U(s)$$

We can separate this into two separate parts:

- $C[sI - A]^{-1}X(0)$ The Zero-Input Response.

- $(C[sI - A]^{-1}B + D)U(s)$ The Zero-State Response.

These are named because if there is no input to the system (zero-input), then the output is the response of the system to the initial system state. If there is no state to the system, then the output is the response of the system to the system input. The complete response is the sum of the system with no input, and the input with no state.

Discrete MIMO Systems

In the discrete case, we end up with similar equations, except that the $X(0)$ initial conditions term is preceded by an additional z variable:

$$X(z) = [zI - A]^{-1}zX(0) + [zI - A]^{-1}BU(z)$$
$$Y(z) = C[zI - A]^{-1}zX(0) + C[zI - A]^{-1}BU(z) + DU(z)$$

If $X(0)$ is zero, that term drops out, and we can derive a Transfer Function Matrix in the Z domain as well:

$$Y(z) = (C[zI - A]^{-1}B + D)U(z)$$

[Transfer Matrix]

$$(C[zI - A]^{-1}B + D = H(z)$$

[Transfer Matrix Description]

$$Y(z) = H(z)U(z)$$

Example: Pulse Response

For digital systems, it is frequently a good idea to write the pulse response equation, from the state-space equations:

$$x[k + 1] = Ax[k] + Bu[k]$$
$$y[k] = Cx[k] + Du[k]$$

We can combine these two equations into a single difference equation using the coefficient matrices A, B, C, and D. To do this, we find the ratio of the system output vector, $Y[n]$, to the system input vector, $U[n]$:

$$\frac{Y(z)}{U(z)} = H(z) = C(zI - A)^{-1}B + D$$

So the system response to a digital system can be derived from the pulse response equation by:

$$Y(z) = H(z)U(z)$$

And we can set $U(z)$ to a step input through the following Z transform:

$$u(t) \Leftrightarrow U(z) = \frac{z}{z-1}$$

Plugging this into our pulse response we get our step response:

$$Y(z) = (C(zI - A)^{-1}B + D)\left(\frac{z}{z-1}\right)$$

$$Y(z) = H(z)\left(\frac{z}{z-1}\right)$$

REALIZATIONS

Realization

Realization is the process of taking a mathematical model of a system (either in the Laplace domain or the State-Space domain), and creating a physical system. Some systems are not realizable.

An important point to keep in mind is that the Laplace domain representation, and the state-space representations are equivalent, and both representations describe the same physical systems. We want, therefore, a way to convert between the two representations, because each one is well suited for particular methods of analysis.

The state-space representation, for instance, is preferable when it comes time to move the system design from the drawing board to a constructed physical device. For that reason, we call the process of converting a system from the Laplace representation to the state-space representation "realization".

Realization Conditions

- A transfer function G(s) is realizable if and only if the system can be described by a finite-dimensional state-space equation.
- (A B C D), an ordered set of the four system matrices, is called a realization of the system G(s). If the system can be expressed as such an ordered quadruple, the system is realizable.
- A system G is realizable if and only if the transfer matrix G(s) is a proper rational matrix. In other words, every entry in the matrixG(s) (only 1 for SISO systems) is a rational polynomial, and if the degree of the denominator is higher or equal to the degree of the numerator.

Realizing the Transfer Matrix

We can decompose a transfer matrix $G(s)$ into a *strictly proper* transfer matrix:

$$G(s) = G(\infty) + G_{sp}(s)$$

Where $G_{sp}(s)$ is a strictly proper transfer matrix. Also, we can use this to find the value of our D matrix:

$$D = G(\infty)$$

We can define $d(s)$ to be the lowest common denominator polynomial of all the entries in $G(s)$:

Remember, q is the number of inputs, p is the number of internal system states, and r is the number of outputs.

$$d(s) = s^r + a_1 s^{r-1} + \cdots + a_{r-1}s + a_r$$

Then we can define G_{sp} as:

$$G_{sp}(s) = \frac{1}{d(s)} N(s)$$

Where

$$N(s) = N_1 s^{r-1} + \cdots + N_{r-1}s + N_r$$

And the N_i are $p \times q$ constant matrices.

If we remember our method for converting a transfer function to a state-space equation, we can follow the same general method, except that the new matrix A will be a block matrix, where each block is the size of the transfer matrix:

$$A = \begin{bmatrix} -a_1 I_p & -a_2 I_p & \cdots & -a_{r-1} I_p & -a_r I_p \\ I_p & 0 & \cdots & 0 & 0 \\ 0 & I_p & \cdots & 0 & 0 \\ \vdots & \vdots & \ddots & \vdots & \vdots \\ 0 & 0 & \cdots & I_p & 0 \end{bmatrix}$$

$$B = \begin{bmatrix} I_p \\ 0 \\ 0 \\ \vdots \\ 0 \end{bmatrix}$$

$$C = \begin{bmatrix} I_p & 0 & 0 & \cdots & 0 \end{bmatrix}$$

Chapter 5

CONTROLLERS AND COMPENSATORS

CONTROLLABILITY AND OBSERVABILITY

System Interaction

In the world of control engineering, there are a slew of systems available that need to be controlled. The task of a control engineer is to design controller and compensator units to interact with these pre-existing systems. However, some systems simply cannot be controlled (or, more often, cannot be controlled in specific ways). The concept of controllability refers to the ability of a controller to arbitrarily alter the functionality of the system plant.

The state-variable of a system, x, represents the internal workings of the system that can be separate from the regular input-output relationship of the system. This also needs to be measured, or *observed*. The term observability describes whether the internal state variables of the system can be externally measured.

Controllability

Complete state controllability (or simply controllability if no other context is given) describes the ability of an external input to move the internal state of a system from any initial state to any other final state in a finite time interval

We will start off with the definitions of the term controllability, and the related term reachability.

Controllability

A system with internal state vector x is called controllable if and only if the system states can be changed by changing the system input.

Reachability

A particular state x_1 is called *reachable* if there exists an input that transfers the state of the system from the initial state x_0 to x_1 in some finite time interval $[t_0, t)$.

We can also write out the definition of reachability more precisely:

A state x_1 is called reachable at time t_1 if for some finite initial time t_0 there exists an input u(t) that transfers the state x(t) from the origin at t_0 to x_1.

A system is reachable at time t_1 if every state x_1 in the state-space is reachable at time t_1.

Similarly, we can more precisely define the concept of controllability:

A state x_0 is controllable at time t_0 if for some finite time t_1 there exists an input u(t) that transfers the state x(t) from x_0 to the origin at time t_1.

A system is called controllable at time t_0 if every state x_0 in the state-space is controllable.

Controllability Matrix

For LTI (linear time-invariant) systems, a system is reachable if and only if its controllability matrix, ζ, has a full row rank of p, where p is the dimension of the matrix A, and $p \times q$ is the dimension of matrix B.

[Controllability Matrix]

$$\zeta = [B \quad AB \quad A^2B \quad ... \quad A^{p-1}B] \in R^{p \times pq}$$

A system is controllable or "Controllable to the origin" when any state x_1 can be driven to the zero state $x = 0$ in a finite number of steps.

A system is controllable when the rank of the system matrix A is p, and the rank of the controllability matrix is equal to:

$$Rand(\zeta) = Rank\ (A^{-1}\ \zeta) = p$$

If the second equation is not satisfied, the system is not.

MATLAB allows one to easily create the controllability matrix with the *ctrb* command. To create the controllabilty matrix ζ simply type

$$\zeta = ctrb\ (A,\ B)$$

where A and B are mentioned above. Then in order to determine if the system is controllable or not one can use the rank command to determine if it has full rank.

If

Rank (A) < p

Then controllability does not imply reachability.

- Reachability always implies controllability.
- Controllability only implies reachability when the state transition matrix is nonsingular.

Determining Reachability

There are four methods that can be used to determine if a system is reachable or not:

1. If the p rows of $\phi(t, \tau) B(t)$ are linearly independent over the field of complex numbers. That is, if the rank of the product of those two matrices is equal to p for all values of t and τ

2. If the rank of the controllability matrix is the same as the rank of the system matrix A.

3. If the rank of rank$[\lambda I - A, B] = p$ for all eigenvalues λ of the matrix A.

4. If the rank of the reachability gramian is equal to the rank of the system matrix A.

Each one of these conditions is both necessary and sufficient. If any one test fails, all the tests will fail, and the system is not reachable. If any test is positive, then all the tests will be positive, and the system is reachable.

Gramians

Gramians are complicated mathematical functions that can be used to determine specific things about a system. For instance, we can use gramians to determine whether a system is controllable or reachable. Gramians, because they are more complicated than other methods, are typically only used when other methods of analyzing a system fail (or are too difficult).

All the gramians presented on this page are all matrices with dimension $p \times p$ (the same size as the system matrix A).

All the gramians presented here will be described using the general case of Linear time-variant systems. To change these into LTI (time-invariant equations), the following substitutions can be used:

$$\phi(t, \tau) \rightarrow e^{A(t-\tau)}$$

$$\phi'(t, \tau) \rightarrow e^{A'(t-\tau)}$$

Where we are using the notation X' to denote the transpose of a matrix X (as opposed to the traditional notation X^T).

Reachability Gramian

We can define the reachability gramian as the following integral:

[Reachability Gramian]

$$W_r(t_0, t_1) = \int_{t_0}^{t_1} \phi(t_1, \tau) B(\tau) B'(\tau) \phi'(t_1, \tau) d\tau$$

The system is reachable if the rank of the reachability gramian is the same as the rank of the system matrix:

rank$(W_r) = p$

<chemistry>/control{range}

Controllability Gramian

We can define the controllability gramian of a system (A, B) as:

[Controllability Gramian]

$$W_c(t_0, t_1) = \int_{t_0}^{t_1} \phi(t_0, \tau) B(\tau) B'(\tau) \phi'(t_0, \tau) d\tau$$

The system is controllable if the rank of the controllability gramian is the same as the rank of the system matrix:

$$\text{rank}(W_c) = p$$

If the system is time-invariant, there are two important points to be made. First, the reachability gramian and the controllability gramian reduce to be the same equation. Therefore, for LTI systems, if we have found one gramian, then we automatically know both gramians. Second, the controllability gramian can also be found as the solution to the following Lyapunov equation:

$$AW_c + W_c A' = -BB'$$

Many software packages, notably MATLAB, have functions to solve the Lyapunov equation. By using this last relation, we can also solve for the controllability gramian using these existing functions.

Observability

The state-variables of a system might not be able to be measured for any of the following reasons:

1. The location of the particular state variable might not be physically accessible (a capacitor or a spring, for instance).
2. There are no appropriate instruments to measure the state variable, or the state-variable might be measured in units for which there does not exist any measurement device.
3. The state-variable is a derived "dummy" variable that has no physical meaning.

If things cannot be directly observed, for any of the reasons above, it can be necessary to calculate or estimate the values of the internal state variables, using only the input/output relation of the system, and the output history of the system from the starting time. In other words, we must ask whether or not it is possible to determine what the inside of the system (the internal system states) is like, by only observing the outside performance of the system (input and output)? We can provide the following formal definition of mathematical observability:

Observability

A system with an initial state, $x(t_0)$ is observable if and only if the value of the initial state can be determined from the system output $y(t)$ that has been observed through the time interval $t_0 < t < t_f$. If the initial state cannot be so determined, the system is unobservable.

Complete Observability

A system is said to be completely observable if all the possible initial states of the system can be observed. Systems that fail this criteria are said to be unobservable.

Detectability

A system is Detectable if all states that cannot be observed decay to zero asymptotically.

Constructability

A system is constructable if the present state of the system can be determined from the present and past outputs and inputs to the system. If a system is observable, then it is also constructable. The relationship does not work the other way around.

A system state x_i is unobservable at a given time t_i if the zero-input response of the system is zero for all time t. If a system is observable, then the only state that produces a zero output for all time is the zero state. We can use this concept to define the term state-observability.

State-Observability

A system is completely state-observable at time t_0 or the pair (A, C) is observable at t_0 if the only state that is unobservable at t_0 is the zero state x = 0.

Constructability

A state x is unconstructable at a time t_1 if for every finite time $t < t_1$ the zero input response of the system is zero for all time t.

A system is completely state constructable at time t_1 if the only state x that is unconstructable at t_0 is x = 0.

If a system is observable at an initial time t_0, then it is constructable at some time $t > t_0$, if it is constructable at t_1.

Observability Matrix

The observability of the system is dependant only on the system states and the system output, so we can simplify our state equations to remove the input terms:

Matrix Dimensions:

A: $p \times p$

B: $p \times q$

C: $r \times p$

D: $r \times q$

$x'(t) = Ax(t)$

$y(t) = Cx(t)$

Therefore, we can show that the observability of the system is dependant only on the coefficient matrices A and C. We can show precisely how to determine whether a system is observable, using only these two matrices. If we have the observability matrix Q:

[Observability Matrix]

$$Q = \begin{bmatrix} C \\ CA \\ CA^2 \\ \vdots \\ CA^{p-1} \end{bmatrix}$$

we can show that the system is observable if and only if the Q matrix has a rank of p. Notice that the Q matrix has the dimensions $pr \times p$.

MATLAB allows one to easily create the observability matrix with the obsv command. To create the observabilty matrix Q simply type

$$Q = obsv(A,C)$$

where A and C are mentioned above. Then in order to determine if the system is observable or not one can use the rank command to determine if it has full rank.

Observability Gramian

We can define an observability gramian as:

[Observability Gramian]

$$W_o(t_0, t_1) = \int_{t_0}^{t_1} \phi'(\tau, t_0) C'(\tau) C(\tau) \phi(\tau, t_0) d\tau$$

A system is completely state observable at time $t_0 < t < t_1$ if and only if the rank of the observability gramian is equal to the size p of the system matrix A.

If the system (A, B, C, D) is time-invariant, we can construct the observability gramian as the solution to the Lyapunov equation:

$$A'W_o + W_o A = -C'C$$

Constructability Gramian

We can define a constructability gramian as:

[Constructability Gramian]

$$W_{cn}(t_0,t_1) = \int_{t_0}^{t_1} \phi'(\tau, t_1) \, C'(\tau) \, C(\tau) \, \phi(\tau, t_1) \, d\tau$$

A system is completely state observable at an initial time t_0 if and only if there exists a finite t_1 such that:

$$\text{rank}(W_0) = \text{rank}(W_{cn}) = p$$

Notice that the constructability and observability gramians are very similar, and typically they can both be calculated at the same time, only substituting in different values into the state-transition matrix.

Duality Principle

The concepts of controllability and observability are very similar. In fact, there is a concrete relationship between the two. We can say that a system (A, B) is controllable if and only if the system (A', C, B', D) is observable. This fact can be proven by plugging A' in for A, and B' in for C into the observability Gramian. The resulting equation will exactly mirror the formula for the controllability gramian, implying that the two results are the same.

SYSTEM SPECIFICATIONS

System Specification

There are a number of different specifications that might need to be met by a new system design. In this chapter we will talk about some of the specifications that systems use, and some of the ways that engineers analyze and quantify systems.

STEADY-STATE ACCURACY

Sensitivity

The sensitivity of a system is a parameter that is specified in terms of a given output and a given input. The sensitivity measures how much change is caused in the output by small changes to the reference input. Sensitive systems have very large changes in output in response to small changes in the input. The sensitivity of system H to input X is denoted as:

$$S_H^X(s)$$

Disturbance Rejection

All physically-realized systems have to deal with a certain amount of noise and disturbance. The ability of a system to ignore the noise is known as the disturbance rejection of the system.

Control Effort

The control effort is the amount of energy or power necessary for the controller to perform its duty.

CONTROLLERS AND COMPENSATORS

Controllers

There are a number of different standard types of control systems that have been studied extensively. These controllers, specifically the P, PD, PI, and PID controllers are very common in the production of physical systems, but as we will see they each carry several drawbacks.

Proportional Controllers

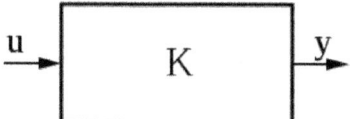

A Proportional controller block diagram

Proportional controllers are simply gain values. These are essentially multiplicative coefficients, usually denoted with a K. A P controller can only force the system poles to a spot on the system's root locus. A P controller cannot be used for arbitrary pole placement.

We refer to this kind of controller by a number of different names: proportional controller, gain, and zeroth-order controller.

Derivative Controllers

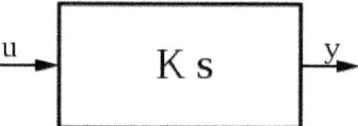

A Proportional-Derivative controller block diagram

In the Laplace domain, we can show the derivative of a signal using the following notation:

$$D(s) = \mathcal{L}\{f'(t)\} = sF(s) - f(0)$$

Since most systems that we are considering have zero initial condition, this simplifies to:

$$D(s) = \mathcal{L}\{f'(t)\} = sF(s)$$

The derivative controllers are implemented to account for future values, by taking the derivative, and controlling based on where the signal is going to be in the future. Derivative controllers should be used with care, because even small amount of high-frequency noise can cause very large derivatives, which appear like amplified noise. Also, derivative controllers are difficult to implement perfectly in hardware or software, so frequently solutions involving only integral controllers or proportional controllers are preferred over using derivative controllers.

Notice that derivative controllers are not proper systems, in that the order of the numerator of the system is greater than the order of the denominator of the system. This quality of being a non-proper system also makes certain mathematical analysis of these systems difficult.

Z-Domain Derivatives

We won't derive this equation here, but suffice it to say that the following equation in the Z-domain performs the same function as the Laplace-domain derivative:

$$D(z) = \frac{z-1}{Tz}$$

Where T is the sampling time of the signal.

Integral Controllers

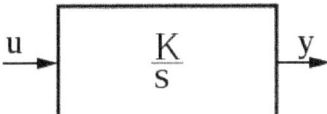

A Proportional-Integral Controller block diagram

To implemenent an Integral in a Laplace domain transfer function, we use the following:

$$\mathcal{L}\left\{\int_0^t f(t)\, dt\right\} = \frac{1}{s}F(s)$$

Integral controllers of this type add up the area under the curve for past time. In this manner, a PI controller (and eventually a PID) can take account of the past performance of the controller, and correct based on past errors.

Z-Domain Integral

The integral controller can be implemented in the Z domain using the following equation:

$$D(z) = \frac{z+1}{z-1}$$

PID Controllers

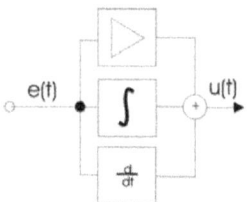

A block diagram of a PID controller

PID controllers are combinations of the proportional, derivative, and integral controllers. Because of this, PID controllers have large amounts of flexibility.

PID Transfer Function

The transfer function for a standard PID controller is an addition of the Proportional, the Integral, and the Differential controller transfer functions (hence the name, PID). Also, we give each term a gain constant, to control the weight that each factor has on the final output:

[PID]

$$D(s) = K_p + \frac{K_i}{s} + K_d s$$

Notice that we can write the transfer function of a PID controller in a slightly different way:

$$D(s) = \frac{A_0 + A_1 s}{B_0 + B_1 s}$$

This form of the equation will be especially useful to us when we look at polynomial design.

PID Signal Flow Diagram

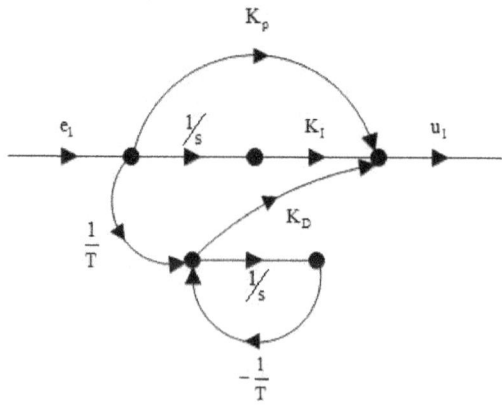

PID Tuning

The process of selecting the various coefficient values to make a PID controller perform correctly is called PID Tuning. There are a number of different methods for determining these values:

1) Direct Synthesis (DS) method
2) Internal Model Control (IMC) method
3) Controller tuning relations
4) Frequency response techniques
5) Computer simulation
6) On-line tuning after the control system is installed
7) Trial and error

Digital PID

In the Z domain, the PID controller has the following transfer function:

[Digital PID]

$$D(z) = K_p + K_i \frac{T}{2}\left[\frac{z+1}{z-1}\right] + K_d\left[\frac{z-1}{Tz}\right]$$

And we can convert this into a canonical equation by manipulating the above equation to obtain:

$$D(z) = \frac{a_0 + a_1 z^{-1} + a_2 z^{-2}}{1 + b_1 z^{-1} + b_2 z^{-2}}$$

Where:

$$a_0 = K_p + \frac{K_i T}{2} + \frac{K_d}{T}$$

$$a_1 = K_p + \frac{K_i T}{2} + \frac{-2K_d}{T}$$

$$a_2 = \frac{K_d}{T}$$

$$b_1 = -1$$

$$b_2 = 0$$

Once we have the Z-domain transfer function of the PID controller, we can convert it into the digital time domain:

$$y[n] = x[n]a_0 + x[n-1]a_1 + x[n-2]a_2 - y[n-1]b_1 - y[n-2]b_2$$

And finally, from this difference equation, we can create a digital filter structure to implement the PID.

Bang-Bang Controllers

Despite the low-brow sounding name of the Bang-Bang controller, it is a very useful tool that is only really available using digital methods. A better name perhaps for a bang-bang controller is an on/off controller, where a digital system makes decisions based on target and threshold values, and decides whether to turn the controller on and off. Bang-bang controllers are a non-linear style of control that this book might consider.

Consider the example of a household furnace. The oil in a furnace burns at a specific temperature -- it can't burn hotter or cooler. To control the temperature in your house then, the thermostat control unit decides when to turn the furnace on, and when to turn the furnace off. This on/off control scheme is a bang-bang controller.

Compensation

There are a number of different compensation units that can be employed to help fix certain system metrics that are outside of a proper operating range. Most commonly, the phase characteristics are in need of compensation, especially if the magnitude response is to remain constant.

Phase Compensation

Occasionally, it is necessary to alter the phase characteristics of a given system, without altering the magnitude characteristics. To do this, we need to alter the frequency response in such a way that the phase response is altered, but the magnitude response is not altered. To do this, we implement a special variety of controllers known as phase compensators. They are called compensators because they help to improve the phase response of the system.

There are two general types of compensators: Lead Compensators, and Lag Compensators. If we combine the two types, we can get a special Lead-Lag Compensator system.

When designing and implementing a phase compensator, it is important to analyze the effects on the gain and phase margins of the system, to ensure that compensation doesn't cause the system to become unstable. Phase lead compensation:- 1 it is same as addition of zero to open loop TF since from pole zero point of view zero is nearer to origin than pole hence effect of zero dominant.

Phase Lead

The transfer function for a lead-compensator is as follows:

[Lead Compensator]

$$T_{lead}(s) = \frac{s-z}{s-p}$$

To make the compensator work correctly, the following property must be satisfied:

$$|z| < |p|$$

And both the pole and zero location should be close to the origin, in the LHP. Because there is only one pole and one zero, they both should be located on the real axis.

Phase lead compensators help to shift the poles of the transfer function to the left, which is beneficial for stability purposes.

Phase Lag

The transfer function for a lag compensator is the same as the lead-compensator, and is as follows:

[Lag Compensator]

$$T_{lag}(s) = \frac{s - z}{s - p}$$

However, in the lag compensator, the location of the pole and zero should be swapped:

$$|p| < |z|$$

Both the pole and the zero should be close to the origin, on the real axis.

The Phase lag compensator helps to improve the steady-state error of the system. The poles of the lag compensator should be very close together to help prevent the poles of the system from shifting right, and therefore reducing system stability.

Phase Lead-Lag

The transfer function of a lead-lag compensator is simply a multiplication of the lead and lag compensator transfer functions, and is given as:

[Lead-Lag Compensator]

$$T_{lead-lag}(s) = \frac{(s - z_1)(s - z_2)}{(s - p_1)(s - p_2)}.$$

Where typically the following relationship must hold true:

$$|p_1| > |z_1| > |z_2| > |p_2|$$

Chapter 6

CONTROLLER DESIGN AND SYSTEM MODELLING

The purpose of this part of the course is to show how settings for controllers can be obtained from a knowledge of the process to be controlled. This forms part of the complete control system design procedure. After manipulated and adjusted quantities have been selected and their pairings, perhaps tentatively, chosen, then values of one or more parameters for each controller must be determined.

The process with these control loops and controller settings can then be tested, usually by simulation using a mathematical model of the process, but sometimes with the 'real' process if it is available. The choice of control loops and/or the controller settings may then be changed if their performance is not satisfactory.

There are three topics to be covered:

- What functional relations or algorithms are to be used in a feedback controller to relate measured error observed to adjustment made?
- How do we obtain, either experimentally or from first principles, a model of a process?
- For a process represented by a given model, what controller parameter values should be used?

CONTROLLER ACTION MATHEMATICS

The basic type of controller is the **Feedback Controller**.

Feedforward controllers also exist but are more complicated to implement. Here we will describe the use of feedback controllers.

In feedback control the variable required to be controlled is measured. This measurement is compared with a given setpoint. The controller takes this error and decides what action should be taken by the manipulated variable to compensate for and hence remove the error.

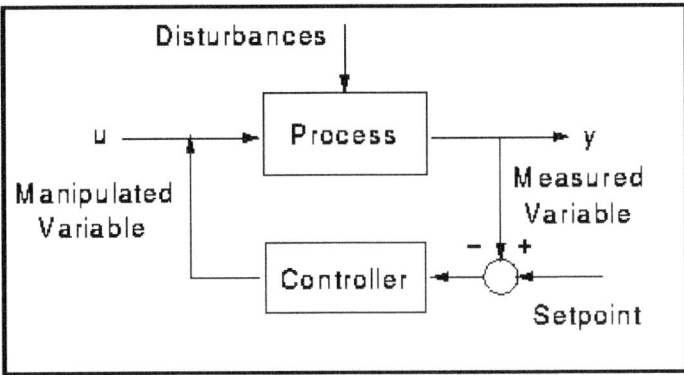

Fig. : Feedback Control Loop.

The **advantage** of this type of control is that it is simple to implement. Not only does the feedback control system require no knowledge of the source or nature of the disturbances, but it also requires minimal detailed information about how the process itself works. Feedback control action is entirely empirical. So long as an adjustment is being made in the correct sense then the control system should remove the effect of an external disturbance.

The **disadvantage** is that the disturbance has to enter and upset the system before it is eliminated.

A feedback control loop can have one of two objectives.

- A servo control loop is one which responds to a change in setpoint. The setpoint may be changed as a function of time (typical of this are batch processes), and therefore the controlled variable must follow the setpoint.

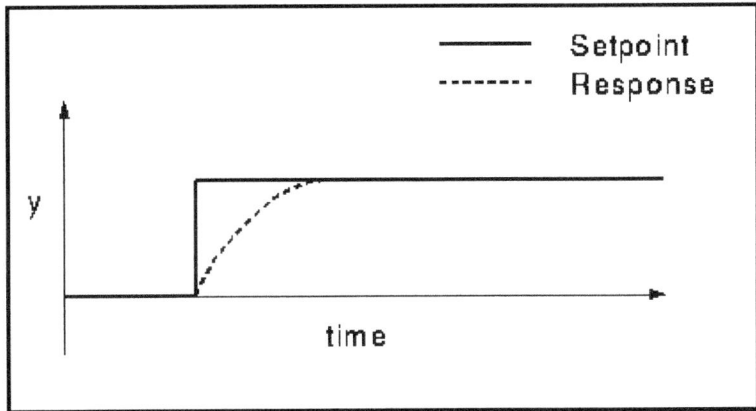

Fig. : Servo Control.

- A regulatory control loop is one which responds to a change in some input value, bringing the system back to steady state. Regulatory control is by far more common than servo control in the process industries.

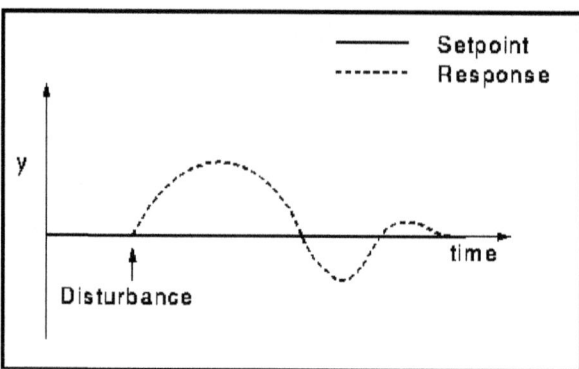

Fig. : Regulatory Control.

Proportional Controller

The first type of controller that we will study is the proportional controller. This controller sets the manipulated variable in proportion to the difference between the setpoint and the measured variable. The bigger the difference, the greater the change in the manipulated variable.

The equation that describes a proportional controller is

$$u_i(t) = \mu(y_s - y(t)) + u_d$$

where

- u_t is the output from the controller, *i.e.* the adjustment
- μ is the constant of proportionality, ususally called the controller gain
- u_d is the output of the controller at its design conditions, sometimes called the bias
- y_s is the required value of y or the setpoint
- y is the input to the controller, *i.e.* the measured variable

The **advantage** of proportional control is that it is relatively easy to implement. However the **disadvantage** is that when implementing a proportional only controller there will be an *offset* in the output. Thus there is always a difference between the setpoint and the actual ouput. The reason why this is so can be shown by means of an example.

Example: Flow Through a Pipe (1)

Below is a diagram of the example.

- F is the flowrate through the pipe
- Fm is the measured flowrate
- Fs is the required, setpoint flowrate
- e is the error between the setpoint and measured value

- Fv is the valve position or controller output

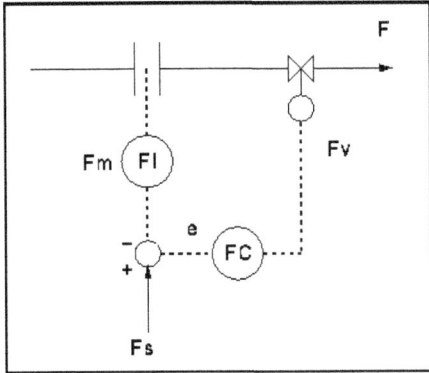

Fig. : Diagram of Flowrate Example.

Therefore we can see that

$$F_m = F = Fv$$

Here

$$y = F_m$$
$$u = F_v$$
$$y_s = F_s$$
$$u_d = F_d$$

So

$$F = \mu(F_s - F) + F_d$$
$$= \frac{(1 + \mu) F_s + (F_d - F_s)}{1 + \mu}$$

Let us assume to begin with that **Fs** = 50 and **Fd** = 50. If this is true then it can be seen from the above equation that **F** = **Fs** and there is no error. Note that this result is independent of the value of the gain.

However, let us now consider what happens when the value of the setpoint changes from 50 to 60 with **Fd** staying constant at 50. First the relevent equation is shown and then the table below summerises the results for different gains.

$$F = \frac{60\mu + 50}{1 + \mu}$$

μ	F	Offset = $F_{sp} - F$
0	50	10
1	55	5
2	57.5	2.5
3	58.75	1.25
10	59.09	0.01

From the above example we can see the problem of using proportional only control, namely the offset. Note also that there must **always** be an offset. This is because to achieve the new steady state the term $\mu(y_s - y)$ **must** have a value and so there **must** be an error. There are two ways of eliminating this problem.

- Choose *ud* to correspond always to the correct output
- Make the gain very large

The first is hard to achieve since it requires very accurate knowledge of the process, and would require changes whenever the setpoint is moved.

The second leads to problems of rangeability and sensitivity. Suppose the gain is 10, then measurement noise of 1% of the total range will cause the control valve to move over 10% of its total travel. This is unacceptable.

Proportional-Integral Controller

To remove the offset integral action is required and so PI control is normally used. It works by summing the current controller error and the integral of all previous errors. It may be thought of as a way of automatically calculating the quantity *ud*. Proper tuning - described in a subsequent section - of the integral part of a PI controller can improve its performance.

If the error *e* is defined as

$$e = y_s - y$$

Then the equation describing a proportional-integral controller is

$$u = \mu(e + \frac{1}{\tau_i} \int edt)$$

where

- τ_i is the reset time of the controller

Alternatively, we can differentiate this expression to get

$$\frac{du}{dt} = \mu \frac{d(y_{sp} - y)}{dt} + \frac{\mu}{\tau_i}(y_{sp} - y)$$

Example - Flow Through a Pipe (2)

Again we will use the example of the flow through a pipe to investigate the nature of Proportional-Integral control.

As before

- $F = y = u$
- $e = Fs - F$

So

$$\frac{dF}{dt} = \mu \frac{d(F_s - F)}{dt} + \frac{\mu}{\tau_i}(F_s - F)$$

Since **Fs** is constant, this becomes

$$(1+\mu)\frac{dF}{dt}=\frac{\mu}{\tau_i}(F_s-F)$$

Or

$$\frac{dF}{dt}=\frac{\mu}{(1+\mu)\,\tau_i}(F_s-F)$$

Equations of the form

$$\frac{dx}{dt}=\frac{1}{T}(a-x)$$

are very common and have well known properties. Their solution has the form shown below.

$$x(t)=a\,(1-\exp\frac{-t}{T})+constant$$

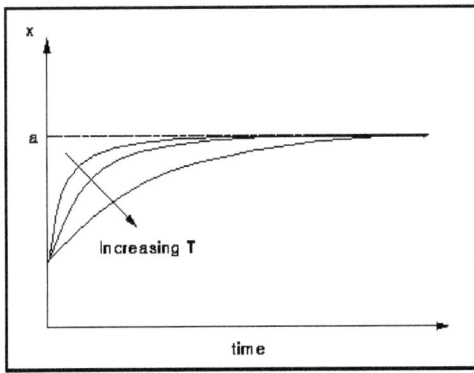

Fig. : Response Curve.

We note that

- As time goes to infinity, $x = a$
- The rate of response of the system increases as T (called the time constant of the equation) becomes smaller

For the flow control system with integral action we see that

- F will eventually become equal to Fs
- This will happen faster if τ_i is small and/or μ is large

MODELLING A PROCESS

There are two ways of approaching the problem of obtaining a mathematical representation or **model** of a chemical process, or indeed anything.

- Create a fundamental or mechanistic model based on knowledge of the physics and chemistry of the system to be modelled. This can be quite a hard thing to do, indeed nearly all of a chemical engineering degree course might be regarded as being about the creation of such models! The advantage of such a model is that it is basically 'right' (provided of course the model builder's knowledge of physics and chemistry is right and is applied correctly). Such a model should be robust in that it can be applied again under conditions of operation different from those for which it was first constructed. If the process being modelled is modified, then analogous modifications to the model will enable it to continue to be used.

We will describe briefly some rules for constructing this type of model which help to ensure that if the modeller's understanding of the problem is correct then a correct model will be obtained.

- Choose a mathematical form which is convenient (*e.g.* it is simple or easy to manipulate) and which represents fairly well the observed behaviour of the system being modelled. Fit numerical parameters to the mathematical form.

This is a so-called 'black box' or 'input-output' model, which seeks only to reproduce the behaviour of the system's output in response to changes in its set-point or inputs. The mathematical form chosen may bear no relation to the form of the equations which truly describe the system. As a result, such models must be used with the greatest care under conditions in the least bit different from those at which the original parameters were determined.

The advantage of such 'arbitrary' models is that they can be developed with little or no knowledge of the system to be represented, and hence complicated systems can be modelled quickly.

Simple Black-Box Models

Input-output models form the basis of most classical process control theory. They are usually subdivided according to whether they have one or more than one input and/or output. We will consider initially only single input, single output (SISO) models, although some ideas associated with multiple input-output models will be touched on elsewhere in the course.

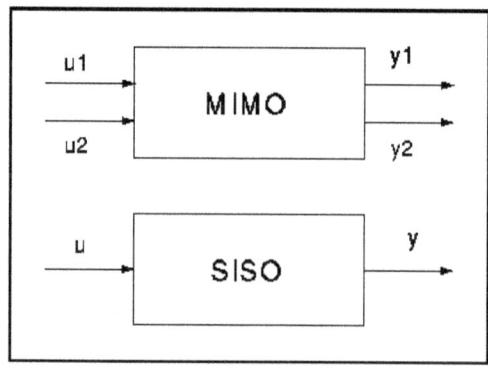

The basic SISO model can be thought of as relating an output y to an input u. In general both of these quanties will change with time, the model must represent how y responds to changes in its input or inputs.

Typical Responses

Suppose an input u is given a **step change** at some time.

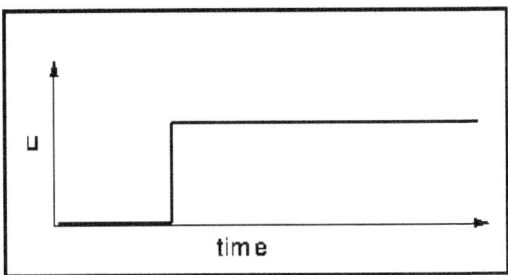

Observations of typical 'processes', from aircraft to papermills, suggest that there are three main types of behaviour which may be seen in an output y.

Instantaneous response

The first typical response is called the instantaneous response. In this case y also responds in a step, but in general of different size to that in u (in any case y will normally have different dimensions to u) as shown below

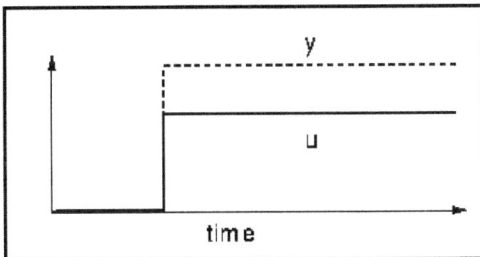

The simplest mathematical relationship is of the form:

$$y(t) = a_1 u(t) + a_0$$

Classical control theory assumes that behaviour can be represented by linear equations like the above, and so this is the only type of equation required to represent this type of behaviour.

In the above equation a_1 is called the **Gain** of the process or model.

Lagging Response

Here y starts to change the moment that y changes, but the full extent of the response 'lags' behind the disturbance. After a while, y will have responded fully.

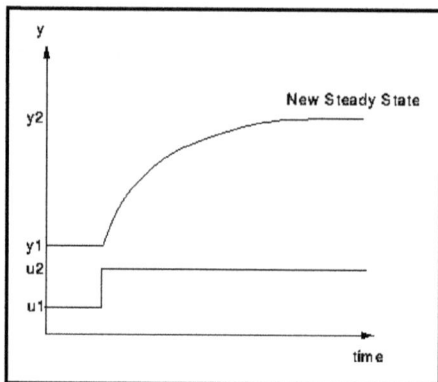

The simplest mathematical form which provides this behaviour is an ordinary differential equation with time t as the independent variable, having the form:

$$\frac{dy(t)}{dt} = \frac{1}{\tau}(a_0 + a_1 u(t) - y(t))$$

Here a_1 is as before the gain, and τ is called the **Time Constant** of the equation, system or model. Because it is described by a single first order o.d.e. this is called a First Order model, system, lag or response. The interpretation of these parameters is described below.

Delayed Response

When u changes, no immediate change in y is observed. However after a time T, y responds completely to the change in u as in the instantaneous response case.

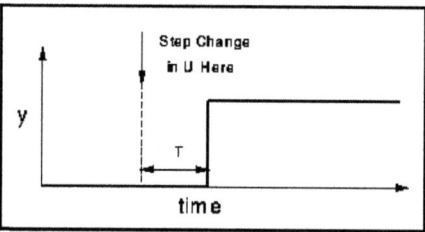

Mathematically this is represented by a difference equation:

$$y(t + T) = u(t)$$

This does not have simple analytical properties, but is easily understood by chemical engineers as corresponding to a plug flow or pipeline system with residence time T. It is also referred to as a time delay or pure time delay system.

Representing Complex Responses

Complex systems may be reasonably well approximated by combinations of the above three elements.

Models of such systems can be assembled as networks of the elements as shown below.

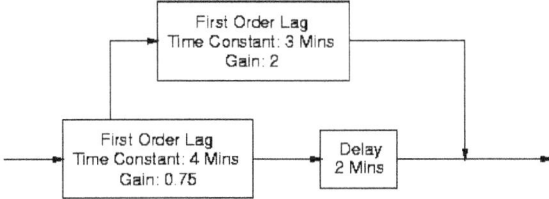

Analytical and numerical techniques are available to work with models constructed in this way.

Theoretical Response

Classical control theory constructs all its models from sets of linear ordinary differential equations. (The instantaneous response is the limiting case of the the o.d.e. where τ is zero, and the plug flow delay, like the plug flow reactor, is the limit of an infinite number of first order lags.)

There is no good physical reason why a real process should be well represented by such a set of equations, except that in the limit of infinitesimally small changes, all nonlinear equations approximate to linear ones.

However, the theoretical advantage of linear representation is twofold. Firstly, the whole system may be represented by o.d.e.s, whereas if there were any nonlinear algebraic equations a mixed set of differential-algebraic equations would be required. Further, a system of linear differential equations always has an analytical solution, but more particularly, is amenable to various other types of analysis which cannot be performed on nonlinear equations. The *tuning* methods for controllers described later make use of this type of analysis to obtain generalised equations for suitable controller settings in terms of parameters of a process model written in terms of the above three types of behaviour. This is not possible for nonlinear systems.

It should be stressed that if we wish to simulate the behaviour of a process, which requires only the solution of the relevant equations, and not their analysis, then there is no particular point in approximating it with this type of simplified approximate model.

Let us look again at the differential equation which describes first order behaviour.

$$\frac{dy(t)}{dt} = \frac{1}{\tau}(a_0 + a_1 u(t) - y(t))$$

It is possible to solve this equation analytically to obtain the expression

$$y = (a_0 + a_1 \Delta u)\left(1 - \exp\frac{-t}{\tau}\right) + y_0 \exp\frac{-t}{\tau}$$

Here

- yo is the value of y at $t = 0$
- Δu is the size of the step change in u at $t = 0$

Note that a graph of this equation gives the response curve shown above under the section on the lag response.

The first thing to consider is What is the Change in y

$$\frac{d\Delta y}{dt} = \frac{1}{\tau}(a_1\Delta u - Dy(t))$$

$$\Delta y(t) = a_1\Delta u(1 - \exp\frac{-t}{\tau})$$

This equation can be now be used directly to calculate the new value of the output variable if the change in u, the gain and time constant are all known. Otherwise it is necessary to estimate values for the gain and time constant as shown below.

Anaysis of Reponse

The tuning controllers that it is useful to be able to look at the open loop response of a process and try and estimate the values of the gain and time constant. Below are notes on how to do this and then you can try it for yourself in the exercises associated with this part of the module.

Estimating the Gain

a_1 is known as the gain. It tells us how much the output variable will change per unit change in the input variable. A large gain implies a large change in y for a given change in u and hence leads to a quicker response.

To calculate its value we have to consider the system going from one steady state value to another. Thus we can see what effect a change in u has on the value of y.

After the system has settled down following the step disturbance

$$\frac{d\Delta y_s}{dt} = 0$$

So

$$a_1\Delta u = \Delta y_s$$

Or, as shown in the graph below

$$a_1 = \frac{\Delta y_s}{\Delta u} = \frac{y_2 - y_1}{u_2 - u_1}$$

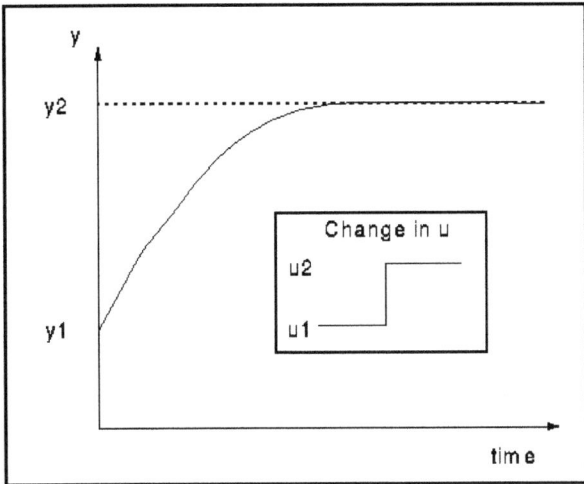

From this we can see that it is a simple calculation to evaluate the gain of the process given the change in *u* and *y*.

Estimating the Time Constant

τ is the time constant for the process. This is related to the speed of response of the system. The diagram below shows a graphical method of evaluating its value.

1. The first stage is to draw the initial slope
2. Then the final steady state value is drawn
3. The time at which these two lines intercept is the value of the time constant

Note that this is also the time taken for the output value to travel 63% of the distance to its new value.

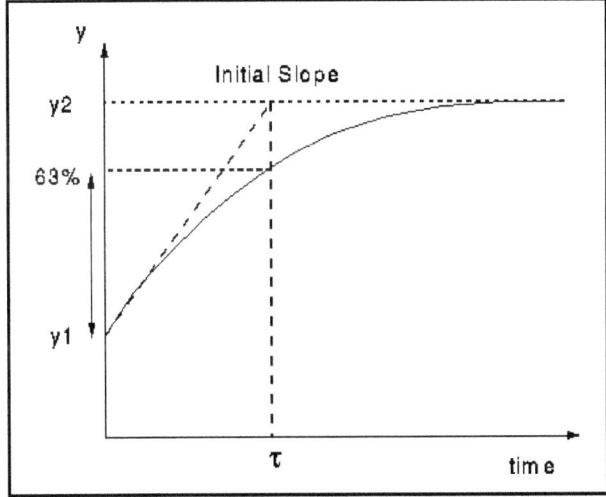

This is shown mathematically below

$$t = \tau$$

$$\Delta y = a_1 \Delta u (1 - \exp(-1))$$

$$= a_1 \Delta u \, 0.63$$

$$\Rightarrow \frac{\Delta y}{\Delta u} = a_1 \, 0.63 = 0.63 \frac{y_2 - y_1}{u_2 - u_1}$$

The following points should be noted about the time constant

- $y(t)$ reaches 63.2% of its final value in one time constant.
- The smaller the time constant the steeper (quicker) the response.
- After 3 to 4 time constants the system is essentially at its new steady state.

Changing the Gain and Time Constant

Finally, how does the response change when a_1 and τ are altered but the change in u stays the same?

The diagram below shows that changing τ alters the slope of the initial slope and changing a_1 alters the final steady state.

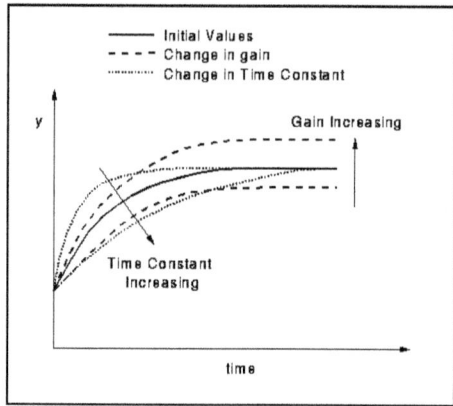

TUNING A CONTROLLER

We looked at two control algorithms for proportional and proportional integral controllers. In order to implement these algorithms there are two parameters which have to be fixed, namely

- μ, the controller gain
- τ_i, the integral reset time

On modelling processes we looked at

- First order input output processes
- Mechanistic models of actual processes

For the first of these we saw that there were three parameters necessary to define the process. These are

- T, the dead time
- a_1, the process gain
- τ, the process time constant

The aim of this chapter is to introduce a method of **matching the personality** of the controller to that of the process so as to achieve the optimum controllability. In other words how do we go from the process parameters to the controller parameters. The method introduced uses the open loop response of a process and works best with a delay-followed-by-first-order-lag. There are many other *tuning* methods which look at other aspects of the process in order to tune the controller.

Selecting Controller Parameters

The **best** choice of controller parameters depends significantly on the nature of the process to be controlled. Thinking back to the simple input-output models we can say that

- Instantaneous Response processes are easy to control. Large gains may be used, subject to noise constraints. Integral action should be used.
- First Order Response Processes are also easy to control. The tuning method described below is based on a first order response.
- Time Delay processes are difficult to control. A pure time delay becomes unstable in principle if a dimensionless gain greater than 1 is used in a proportional only controller.
- Inverse Response processes exhibit a response to an adjustment like this:

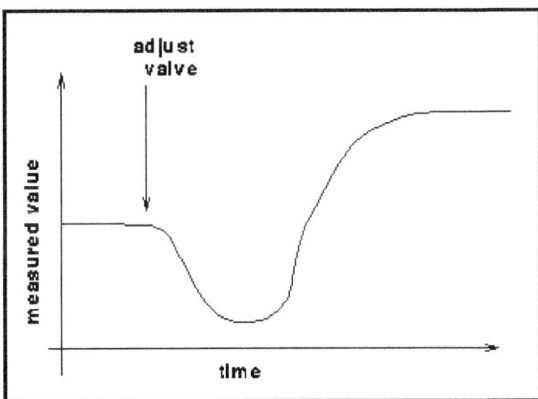

The interactive exercises that this is very difficult to control!

A General Process Model

Let us consider first the simple case of a first-order lag in series with a time delay. This setup is shown in a diagram below.

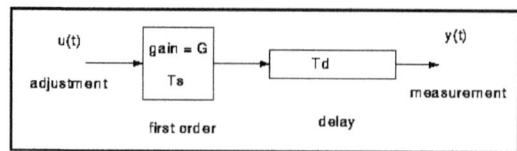

If we change *u* by a known amount and plot the response curve it is possible to determine the model parameters *Ts, Td* and the *gain* from the resulting graph. This is shown in the diagram below.

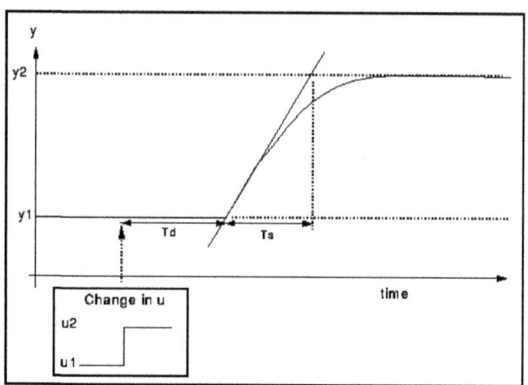

Zeiglar Nichols Open Loop Tuning Method

The Zeigler Nichols Open-Loop Tuning Method is a way of relating the process parameters - delay time, process gain and time constant - to the controller parameters - controller gain and reset time. It has been developed for use on delay-followed-by-first-order-lag processes but can also be adapted to real processes.

The method is outlined below.

- Look at the open loop response of the process to a step change in the manipulated variable.
- Evaluate
  - o The steady-state gain, $(y_2 - y_1) / (u_2 - u_1)$
  - o The time delay, *Td*
  - o The time constant, *Ts*

The diagram above shows how to obtain these values.

- Finally substitute these values into the table below to obtain the relevent controller parameters.

Controller Type	Gain	Reset	Derivative
P	Ts / Td	-	-
PI	0.9 Ts / Td	3.3 Td	-
PID	1.2 Ts / Td	2.0 Td	0.5 Td

The *Gain* evaluated above is the product of the controller gain setting, μ and the process steady state gain, **G**.

Gain $= \mu * G$

Therefore by substituting all the values in for the above and re-arranging we get the following values for the controller parameters:

Controller Type	Controller Gain, μ	Reset	Derivative
P	(Ts Δu) / (Td Δy)	-	-
PI	(0.9 Ts Δu) / (Td Δy)	3.3 Td	-
PID	(1.2 Ts Δu) / (Td Δy)	2.0 Td	0.5 Td

Advantages of this method are:

- Only a single experimental test is needed.
- It does not require trial and error
- The controller settings are easily calculated.

However there are also **Disadvantages**

- Experiment is under open loop response and so disturbances may affect the results.
- Results tend to be oscillatory.
- Does not work well for complex responses - leads to inaccurate tuning model.

Controlling Real Processes

In the real world, unfortunately, the response of a process to a change in one of its inputs seldom follows the first-order case required for the Z-N tuning.

If we are lucky it may be similar in form but different in detail as shown below.

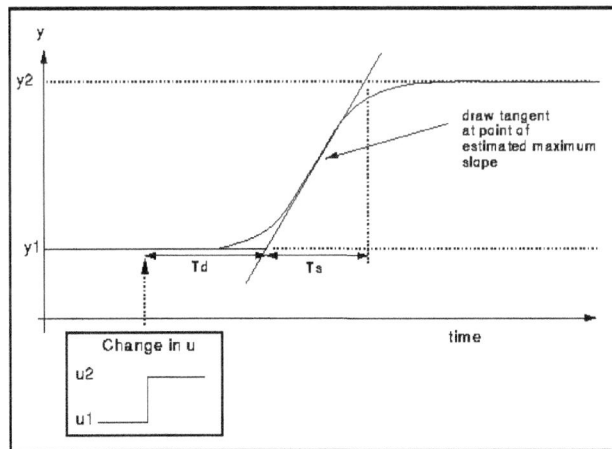

In this case the tangent should be drawn at the point where the slope of the response is steepest. Now we have *estimates* for the parameters and it may be necessary to change them in order to get the optimum values. To do this it is possible to use a model of the process and controller to see the effect of altering the control parameters.

However if we are unlucky the response may be like this...

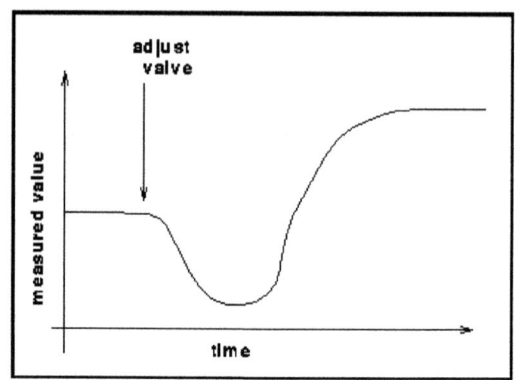

... in which case it is very difficult to control and it may not be possible using a PI controller.

Gains for Real Processes and Controllers

In a process the measurement *y* is strictly speaking a dimensioned quantity: temperature, pressure, flow *etc.*

The adjustment *u* is usually a flow, so that the process gain, $\dfrac{\Delta y}{\Delta u}$, will in general have odd dimensions! This also makes it hard to interpret or compare gain values.

In practice, both measurement and adjustment have a **maximum range** determined by the measuring instrument or valve. It is best to work with **scaled** quantities always expressed as a fraction or percentage of range, *e.g.*

$$\Delta u'(\%) \equiv \frac{100\,\Delta u}{max\ range\ of\ valve}$$

$$\Delta y'(\%) \equiv \frac{100\,\Delta y}{max\ range\ of\ sensor}$$

As indicated these values are percentages and so are *dimensionless* values between 0 and 100. Thus it is possible to define a *dimensionless* gain for the process as

$$G' = \frac{\Delta y'}{\Delta u'}$$

If the value of the gain is large, say 100, then this means that the change in y is 100 times greater than the corresponding change in u. This could lead to y going *out of bounds* or else the change in u being very restrictive.

Alternatively, if the gain is very small, say 0.01, then for a large change in u there is hardly any response in y.

What is required is a gain of around 1. This enables both input and output to be used to their full ranges which in turn improves the controllability.

So if this definition of the gain is used it is clear from a glance if a suitable value has been obtained or not. In this case simply use the value of the *gain* from the first table above along with the dimensionless process gain above to obtain the dimensionless controller gain.

$$gain = \mu' G'$$

In practice a controller does not want to deal with meaningless dimensions when asking for the value of the gain. Therefore a parameter known as the Proportional Band is used instead.

Firstly remember that you have a value of the *dimensionless* gain for the controller as evaluated above.

Now we define the Proportional Band, P, as the reciprocal of the dimensionless controller gain.

$$P = \frac{100\%}{\mu'}$$

Remember that when specifying a controller setting, **always** use dimensionless gain or proportional band.

Chapter 7

CONTROLLING SIMPLE PROCESSES

In all the examples so far discussed it has been assumed that we know at the outset what quantity is to be measured and thus regulated, and what will be the corresponding adjustment. This information will in fact be readily available only for the simplest of cases.

As soon as we consider the control of any sort of process, or even a modestly complex piece of equipment, we are faced with the need to provide *several* control loops. The result of this is to create a range of choices. We introduce ideas to help resolve this choice systematically. These ideas will enable us to design complete control systems for large and complex processes, although they will first be introduced in the context of simple examples.

Degrees of Freedom

When faced with the task of devising a control scheme for a process it is necessary to know *how many of the process variables am I entitled to attempt to regulate*. By process variables we mean temperatures, pressures, compositions, flowrates or component flowrates. The answer arrived at is known as the *number of degrees of freedom*. The degrees of freedom of a process are here defined as the number of process variables which can be set by the designer, operator or control system ie

- Temperatures
- Pressures
- Compositions
- Flowrates (component or total)

In our case we are concerned with the *Control Degrees of Freedom* which will be the number of the above types of process variable which may be set once non-adjustable design variables, such as vessel dimensions or number of trays, have been fixed.

In this context the number of degrees of freedom thus corresponds strictly to the number of manipulated variables which may be used in control loops. Note that this is also the number of single-input-single-output control loops and of regulated variables in the loops.

Example 1: Vapouriser Problem (1)

To illustrate the nature of the problem, consider how a process unit which vaporises a liquid feed stream might be controlled.

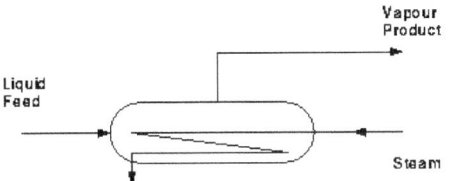

In this device the quantities which we might choose to regulate include:

- Feed rate
- Product rate
- Operating pressure
- Operating temperature
- Liquid level

The first question to be resolved is *which and how many* of these can legitimately be regulated independently?

Similarly, what *adjustments* may be made in order to regulate the chosen quantities? There appear to be three candidates for streams on which control valves might be located, namely:

- Liquid feed
- Vapour product
- Steam supply

Suppose that liquid level is chosen as one of the regulated quantities. Which of these three possible adjustments should be *paired* with this measurement to complete the control loop?

We will return to this particular example after addressing individually the problems noted above. In summary these are:

- How many, and which, quantities can be measured and regulated?
- How many, and which, quantities can be adjusted?
- Which measurement should be paired with which adjustment?

Before proceeding to this it is worth noting the following points about the vaporiser example, to illustrate that these questions can indeed be answered using our knowledge of the process.

- Since there are only three potential adjustments, there cannot be more than three control loops, and hence no more than three regulated quantities.
- Physics and thermodynamics tell us that certain variables in this problem cannot be set independently, and thus cannot be regulated in separate control loops. Unacceptable combinations here are:
  - o Inlet and outlet flows, which must be the same by conservation, and
  - o Temperature and pressure, which are related in a single component two phase system.
- This leaves us with three possible control loops, in which the three adjustments regulate:
  - o Temperature or pressure,
  - o Feed rate or product rate,
  - o Liquid level or holdup.
- The choice of pairing remains, to some extent, as a genuine choice between alternatives which must each be evaluated. Here any of the three possible adjustments can be seen to affect both temperature (or pressure) and holdup, two (vapour and steam valves) can affect vapour rate, and only one can affect feed rate.

Example 2: Mixing of Two Streams

Consider the exceptionally simple process shown below in figure (a), where two feed streams are mixed together to produce a single product stream. Suppose that what is required is that the two feed streams shall have individually specified flowrates. A suitable control system for this would be as shown in figure (b).

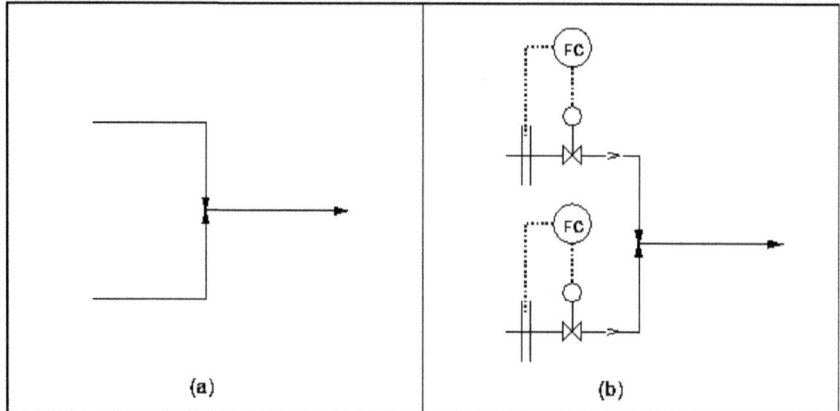

It should be immediately apparent that this control system is complete, *i.e.* we cannot put any more control loops on it. Having fixed two of the three streams which are connected together, conservation requires that the third must be the sum of these two.

This suggest the validity of the following rule.

Rule 1: $(n - 1)$ out of n'. If n streams join together in a process or part of a process over which mass must be conserved (normally *any* process), then the flows of only $(n - 1)$ of these may be set by flow controllers.

It is possible to prove this formally. As will be seen, further generalisation is also possible.

Conservation of mass requires that the sum of the inflow and outflows shall match over an extended period, but is it necessary to take steps to ensure that this happens from minute to minute? The flows would not always match if there were any possibility of material accumulating within the junction. This will not occur if the fluids are incompressible. This implies that it *is* necessary to have some mechanism to ensure that mass balances do actually balance. In this example, the design of the process, *i.e.* simple closed junction and incompressible fluid, ensures that this will be so.

However, consider what happens if we replace the closed junction by an open tank, see below figure (a). Here there is nothing to stop the tank from running dry or overflowing, unless, as in figure (b), we provide the tank with a level controller.

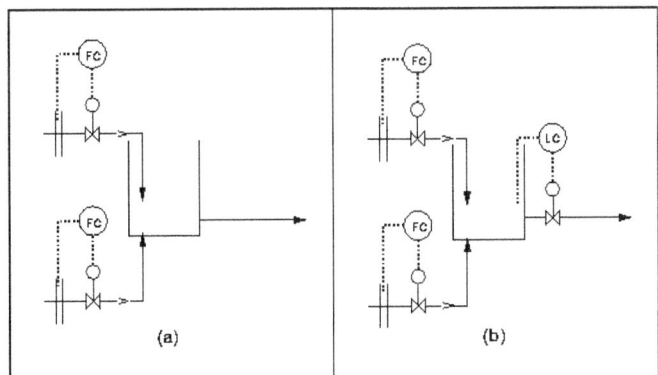

Rule 2: Mass balances must balance. To ensure that mass balances *do* balance, there must either be an inplicit mechanism in the process, or an inventory controller must be supplied. The valve for the holdup control loop goes on the remaining stream.

Inventory or holdup is measured either by level or some equivalent measurement in a liquid system, or by pressure in a gas system. A simple junction with a liquid system is rather special in being 'self regulating' with respect to holdup.

If we wished to regulate the total product rate from a simple mixing process and one only of the feeds, then we may require explicit holdup control for either compressible or incompressible fluids in either of the arrangements below. The vessel in the right hand figure is a closed vessel, which, in the case of a liquid system, would be run full. Whether or not a control loop for pressure is required will depend on the specific process conditions, in particular the source and sink pressures for the flows and the type of device, pump, compressor *etc.*, if any, driving the flow.

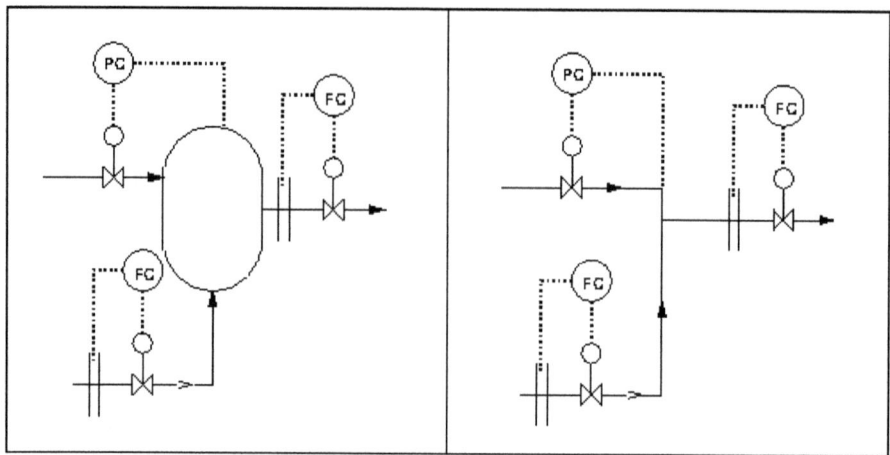

Finally, there is a further rule implied by all these examples which really belongs before any of the others:

Rule 3: Strategic aims. The primary control objectives of a process are set by the strategic aims of the process. These define the basic control structure.

Thus the fact that flow controllers were placed on the two feed streams in the first example and on the product and one feed in the last was a consequence of a decision by the *process* designer that these were the streams whose flows were to be fixed. It should thus be clear that the design of a process and of its control system cannot really be separated.

Hierarchical Decomposition

In the above examples it was simple enough to look at the process or single piece of equipment and see how many degrees of freedom there were and how the control loops would interact. What happens, however, when it is a complete process that is to be controlled with many loops.

The design of a process control system. This approach will be seen to have a number of advantages. Firstly, it provides a systematic approach to resolving what can otherwise seem to be a complex and unstructured problem. Secondly, it enables us to concentrate on individual parts of the problem, rather than trying to do several things at once. Finally, it corresponds to a standard systematic approach to designing *processes*, enabling us to evolve the design of the process and its control system together.

The process and control scheme can be looked in to at differing levels of complexity:

- Input-Output Stage: Strategic Decisions
- Functional Level Stage: Further Details
- Separation Stage: Final Details

There now follows two examples of this hierarchical approach in use. Firstly there is a simple process and then we return to the vapouriser problem introduced earlier.

Controlling a Real Process

Consider a rather more sophisticated mixing process.

Process description, (i): The aim of the process is to deliver a fixed amount of product, made by blending together two streams of two constituents, a concentrate and a diluent, and to supply this product to a specified composition.

In this example we shall introduce a *hierarchical* procedure for developing the design of a process control system. This approach will be seen to have a number of advantages. Firstly, it provides a systematic approach to resolving what can otherwise seem to be a complex and unstructured problem. Secondly, it enables us to concentrate on individual parts of the problem, rather than trying to do several things at once. Finally, it corresponds to a standard systematic approach to designing *processes*, enabling us to evolve the design of the process and its control system together.

Input-Output Stage: Strategic Decisions

Starting from the above statement of the process requirements, viz specified product rate and composition, without reference to any detail of the process itself, other than the *input* and *output* streams, we can define immediately a part of the control system structure, as shown below. Here the, unspecified, process is shown as a box. It is clear that the product stream will require flow control and that that can be implemented as shown. It is also clear that composition measurement and some sort of composition or quality control loop will be required. Half of this loop can be immediately defined, and is also shown.

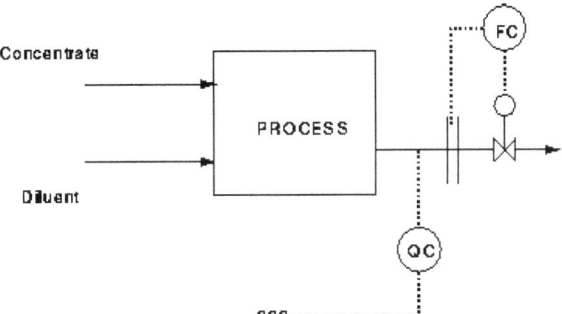

Still without any detailed knowledge of the contents of the PROCESS box, consider how the composition control loop might be implemented. What can be adjusted to cause the composition of the product stream to change? Clearly, it will be necessary to manipulate *either* the amount of diluent *or* the amount of concentrate. These lead to two *alternative* structures shown below.

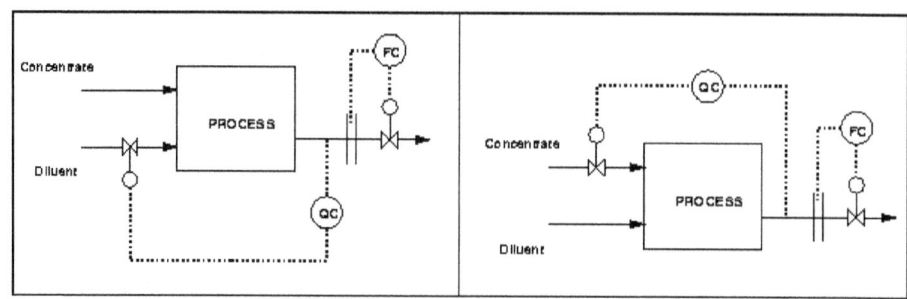

Which of these is the better structure? Without detailed and quantitative information about the process in the box, it is not possible to decide. This is a common situation in engineering design. Ultimately, it may well be necessary to explore both alternatives, and make a decision on the basis of some measure of overall system performance. The designer could proceed with both alternatives in parallel. Unfortunately, this is almost certainly only the first of many points where alternatives arise, and very soon the 'tree' of possible designs will become intractably large. Unless the whole design procedure is automated and carried out by a very powerful computer, this is not a realistic approach.

The following *heuristic* approach, based on the ideas of Douglas is recommended.

- Based on information currently available, choose the more promising alternative using rules-of-thumb. This might mean making an arbitrary choice.
- Note this point in the design as on at which a choice was made.
- Proceed with the chosen alternative until either:
  - o it is found to be unsatisfactory, or
  - o the design is complete.
- If the design was unsatisfactory, backtrack to the decision point and explore the other branch.
- If the design was completed, then evaluate it and:
  - o if satisfied that the design is the best available, or good enough, *finish,* or
  - o if not entirely satisfied, *backtrack* and explore the other branch.

This algorithm is a rather general one for any kind of design.

Here we can apply the following heuristic or rule-of-thumb.

Heuristic : Small streams. Manipulate small streams rather than large ones in important control loops.

This has a number of justifications. Firstly, small valves are cheaper than large ones, so it may be possible to save money. Secondly, small valves can be manipulated more quickly and precisely than large valves, and so a control loop with a smaller valve will often work better.

Clearly, the concentrate stream will be a smaller one than the diluent, and so we will choose to follow up the right hand alternative where this is the manipulated variable for the composition loop.

Examination of the flowsheet shows that we have control valves on two out of the three streams associated with the process. Since these flows have been set, one to a specific flow and another to ensure that a particular product composition is achieved, the flow of the third stream cannot now be chosen independently, it must match these two flows to ensure that the mass balance is maintained. In fact our '$(n-1)$' rule, and its corrollary, can be generalised to cover *any* type of controller with valves on $(n-1)$ out of n streams.

Rule 1a : Generalised '(n-1) out of n'. If n streams join together in a process or part of a process over which mass must be conserved (normally *any* process), then the flows of only $(n - 1)$ of these may be set by control loops other than one regulating inventory within the process or part process.

Without a knowledge of precisely the type of process element on the box we cannot completely define any control loop associated with inventory or holdup regulation, but we do know that we *cannot* put a control valve on the remaining stream for any purpose *other* than inventory regulation. We will indicate this on the flowsheet as shown below. The shaded 'valve' implies that no other valve may be put on this line. The square, rather than round 'controller' indicates that some mechanism, not necessarily an actual controller, will regulate inventory, which might, if measured, be a level, mass holdup or pressure.

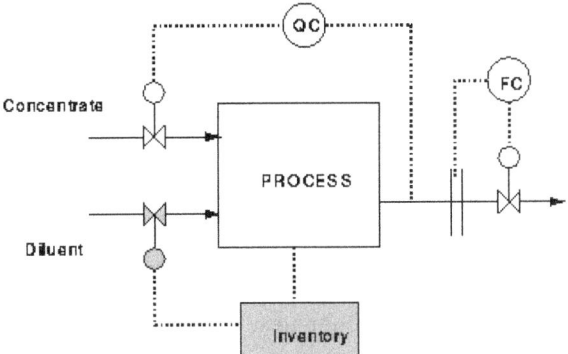

Functional Level: Further Details

The steps which were followed above illustrate that it is sometimes possible to design a significant part of the control system for a process by reference to:

- The objectives of the process with respect to amount and composition of products.
- The basic structure of the process.

It will not in general be possible to determine the whole control system with just this information. Further steps in developing this involving 'opening up' the box labelled PROCESS in this example into succesive levels of increasing detail. This approach will be explored in later examples. The purpose of this *hierarchical* approach is to help the designer to concentrate on the decisions that can be taken at each stage, by presenting details of the process in sequence rather than all at once. This makes the design task easier both by reducing the amount of new information presented at one time, and by allowing some earlier decisions to be finalised and thus removed from the list of tasks still to be tackled. A glance ahead at a complete process flowsheet will enable the reader to appreciate how daunting a task placing the control loops on a flowsheet might be if this approach is *not* adopted.

This example may however be completed in one further level by providing some more details of the process.

Process description, (ii): The process equipment consists of an open mixing tank, followed by an in-line static mixer (a section of pipe with internal vanes or baffles) and a second mixing tank.

The PROCESS box to include this, and with the control system so far defined, is shown below.

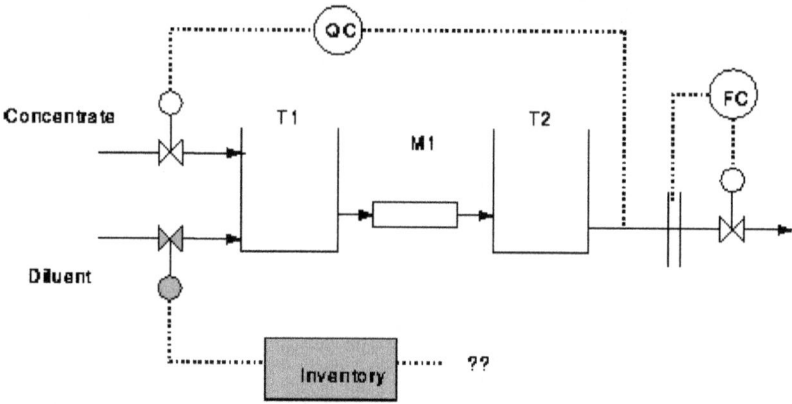

Looking at this more detailed structure, the questions to ask, in order, are:

1. Can any partial control loops now be completed?
2. Look at 'new' streams created by expanding the structure: what *strategic* control requirements arise?
3. Having identified strategic controls, can any inventory controls be identified using the '(n-1)' and mass balance rules?

In answering question 2, it is important to realise that in any process, streams do not simply *happen*! They are their for a reason, and that reason usually defines what will determines their flow.

Refering to process, the incomplete inventory loop cannot be unambiguously completed. Intuition suggests that it should probably regulate the level in T1, and this will indeed prove to be the case, but we cannot be certain of this yet.

The two new streams, from T1 to M1 and from M1 to T2, have no obvious strategic requirements for regulation of flow or composition.

However, T2 has two streams, one of which has its flow set by the product flow control loop. The tank is not self regulating, and so an level controller must be placed with a valve on the feed stream as shown.

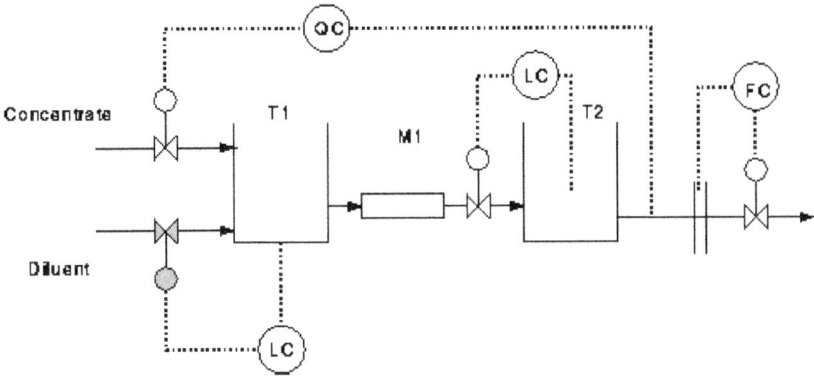

It now becomes clear that mixer M1 requires inventory regulation, having one of its two streams now set, which would imply a valve on its input stream from M1. However, it is self regulating, being essentially just a piece of pipe, but this still means that no valve can be placed there for any other reason. It is now clear that the measurement end of the original level control loop must be in T1.

All streams being accounted for by having explicit control valves or implicit regulatory mechanisms determining their flows, the control scheme is now complete.

Review: What Did We Do?

This was a very simple process. However, there were potentially a significant number of alternative process structure, *not all of which would have worked*. We applied a logical procedure, each of whose steps could be justified with reference either to a knowledge of the process or the rule which have been proposed, and ended up with a complete, and workable, control system with four loops.

How Many and which Quantities to Measure?

The question of 'how many?' was not posed or answered explicitly in this example. As will be seen later, it is sometimes convenient to do so. However in this case it was subsumed in the question of 'which?'.

The question of 'which quantities to measure' was answered in two ways. Firstly by reference to the Strategic aims of the process rule. This established the outlet flow and concentration as regulated quantities. Further strategic regulated variables, other than inventories can often be established by by reference to identified adjustable variables, but in this process there were no others.

Secondary regulated variables are usually inventories and are identified by the $(n - 1)$ and mass balance rules.

How Many and which Quantities to Adjust?

It is clear that there cannot be more adjusted variables than there are streams whose flows can be manipulated independently. These were all identified in this process by the requirement for inventory regulation. In general we can use the following Rule both to identify adjustments and to check the final control system structure.

Rule: flows do not just happen. Stream flows in a process do not just happen. Either a valve or a mechanism (such as continuity) must set the flow of every stream.

Which Measurements and Adjustments are Paired?

Here there will almost invariably be alternatives. To identify these and help choose between them we have one firm Rule, and some guiding heuristics.

Rule: Cause-and-effect. An adjustment chosen to pair with a measurement must have an effect on that measurement.

This is rather obvious. It is nonetheless given a name in the control literature where it is called *structural controllability*.

A number of heuristics serve to aid choice. One has been given, the Small streams heuristic.

Heuristic: immediate response. Prefer pairings in which the measurement responds immediately, rapdicly and unambiguosly to the adjustment.

Their are a number of important quantitative elements embodied in this heuristic. These are dealt with in detail elsewhere.

Heuristic: noninteraction. An ideal adjustment should affect its paired measurment and no other measurements.

Vapouriser Problem (2)

We now have sufficient understanding of how to develop whole process control schemes to tackle the vaporiser problem.

Input-Output Stage: Stategic Decisions

Redrawing this 'process' as an input-output block yields the structure shown. In the block we have distinguished two sub-blocks, noticing that while both feed and steam enter the process they are subject to separate material balances.

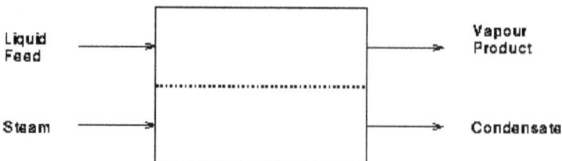

The objectives of this process will be taken to be:

1. deliver a specified quantity of vapour,
2. at specified temperature and pressure.

Objective 1. suggests that we should place a flow controller on the vapour outlet as shown in figure (a). Noting that the process fluid side of the block has only one input and one output, the feed stream may be 'blocked' for any purpose other than inventory regulation as shown by the shaded valve.

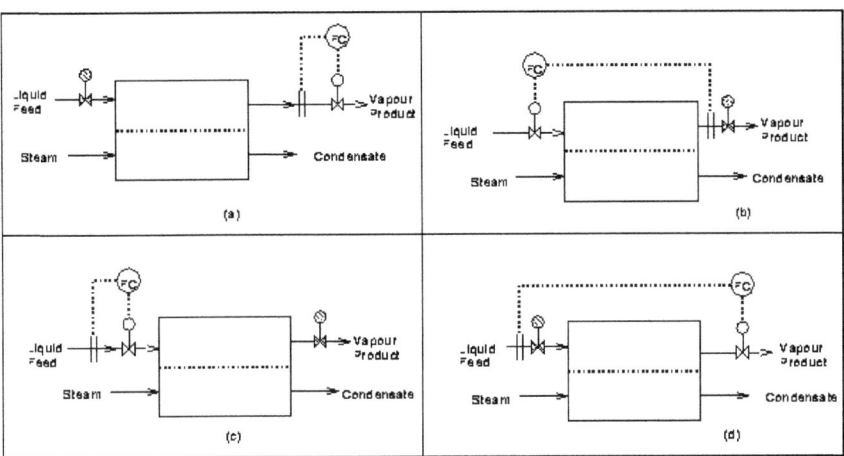

Figure (a) fulfills the stated objective of delivering a specified rate of vapour product. However, because there is only one input and one output to the process side, so would the structure of figure (b). The first structure is prefered according to the following Heuristic.

Heuristic: close adjustment. It is usually better to make an adjustment as close as possible to the measurement with which it is paired.

Also because the sub-block has only two streams, the scheme in figure (c), which regulates the *feed* flow would also serve to maintain the product rate once inventory regulation was provided. This is clearly a less direct way of performing the specified task, and depends of the satisfactory operation of the inventory control system. It is therefore avoided on the basis of a further Heuristic.

Heuristic: direct action. Prefer the the structure which manipulates the regulated quantity most directly. In particular, avoid arrangements which depend on the satisfactory operation of additional control loops.

The final structure, figure (d) violates both of the above guidelines.

Objective 2. explicitly states that the vapour temperature shall be regulated, as shown below, (a). Since all streams on the process side are set when inventory regulation is added, only the steam rate can be adjusted. If this is done on the steam supply, then the condensate outflow must be adjusted to maintain the steam side material balance, figure (b).

This control system must now be complete, as there are no more possible adjustments. The designer cannot therfore be tempted to try and regulate the pressure of the system. The Phase Rule could also have been invoked to check this.

$$N = C - P + 2$$

N is the nuber of extensive thermodynamic variables which may be chosen or set, C is the number of components and P the number of phases. Here $P = 2$, there being both vapour and liquid present and so:

$$n = C$$

However, the feed composition is fixed and because of the requirements for material balance, the vapour will have this same composition. Specifying the composition on a C component stream fixes $(C - 1)$concentrations, leaving one intensive variable which can be fixed by a control system. This can be *either* temperature *or* pressure, but clearly not both independently.

Rule: Phase Rule. The Phase Rules still says:

$$N = C - P + 2$$

Completing the Example

The control system structure has been completely defined at the input-output level. This is rather unusual, but it is now very easy to open up the block and turn the conceptual controllers into 'real' ones, see below figure (a). Note the special form of inventory regulation on the steam heating side using a steam trap, essentially a very small vessel with an internal level control system to allow steam and condensate to be disengaged.

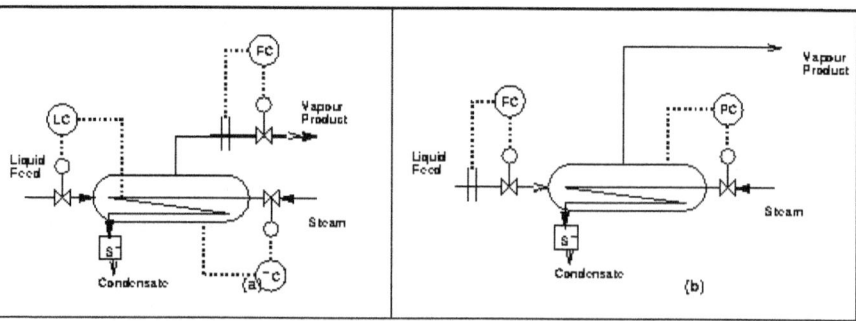

Just to show that no heuristics are totally reliable, figure (b) shows another version of the control system which would be acceptable in many circumstances. This breaks the Direct action heuristic by regulating the feed rather than the product. However, it saves a control loop by having a 'self regulating' material balance through the following mechanism.

Heat transfer from the coil to the fluid happens only in the liquid phase. If too much vapour leaves the vessel, the liquid level falls, uncovering some of the steam tubes. Thus the heat transfer area falls, reducing the rate of heat transfer and the rate of vaporisation. Normal operation will be with the tube bundle partly uncovered.

This arrangement will only be acceptable if this is allowable, *e.g.* with relatively low temperature steam. Another problem could be that there will always be some liquid boiling dry on the exposed tube surface. This could tend to degrade and build up deposits.

Finally, to show that, in this system, temperature and pressure regulation are equivalent, the temperature control loop on the steam has been replaced by a pressure controller.

Rigorous Identification of Number of Regulated Quantities

It is possible to determine the number of variables in a process which it is permisable to regulate by a formal mathematical procedure. In the previous example it this was not necessary, as identification of the number of available adjustments served to define this. However, it will sometimes be the case that two or more potential adjustments have the same effect and are alternatives in the sense that if one is used, the other cannot be. It is also possible, in principle, to devise a process which is uncontrollable because it has more variables to regulate than there are adjustments available.

The mathematical technique is called *degrees of freedom analysis*. It can be carried out in a number of ways, but essentially it consists of counting the number of equations required to describe the process or part process under consideration, and the number of variables which appear in these equations. The excess of variables over equations represents the number of quantities which may be set to arbitrary values, for example by a control system.

Chapter 8

CONTROL OF SEPARATION PROCESSES

This module deals with the control of a couple of simple separation processes:
- Adiabatic Flash
- Non-adiabatic Flash

Adiabatic Flash: Degrees of Freedom

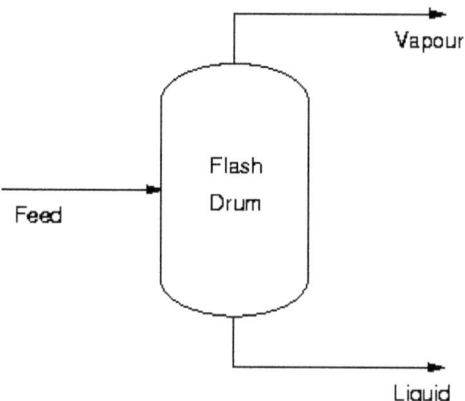

Before we can attempt to control this process we have to know how many streams we are allowed to manipulate *i.e.* the control degrees of freedom. This can be evaluated by the equation

$$d.o.f = unknowns - equations$$

Unknowns

There are 3 streams in this process. Each stream has unknown composition, temperature and pressure. Thus for a stream with C components there are

$$3\ [C + 2] = 3C + 6\ unknowns$$

Equations

In an adiabatic flash there are

- C material balances
- C equilibrium relationships
- 1 energy balance
- 1 equation relating the temperature of the products *i.e.* they must be the same.
- 1 equation relating the pressure of the products *i.e.* they must also be the same.

Hence altogether there are

$$2C + 3\ constraints$$

Degrees of Freedom

From the above we can evaluate that we have

$$C + 3\ degrees\ of\ freedom$$

However we know the composition, temperature and pressure of the input stream. These were considered unknowns for the above calculations but now can be taken into account. Thus this adds on another $C + 1$ constraints and so we now have

$$2\ control\ degrees\ of\ freedom\ or\ C.D.F$$

This answer can be compared with that obtained directly using the equation below.

$$C.D.F = no.\ of\ connections + 1 - no.\ of\ phases$$

- Number of connections for an adiabatic flash = 3
- Number of phases = 2
- Hence C.D.F = 3 + 1 - 2 = 2, as expected!

Adiabatic Flash: Control Strategy

It is now possible to devise control schemes for this adiabatic flash vessel. The basic input-output block has 3 streams. It has been evaluated above that there are 2 control degrees of freedom which means control valves on two of the three streams. That leaves one other stream which also has a control valve on it to regulate inventory *i.e.* to ensure that the mass balance does in fact balance.

Of the two streams left one would normally have a flow controller and be used to regulate throughput. This could be any of the three streams. This leaves one other to regulate a *strategic* variable.

From a knowledge of the properties of the adiabatic flash, the normal design specification is the feed and only one further quantity, usually pressure, but temperature or another flow could also be chosen.

Adiabatic Flash: Examples

For the examples below we will choose to regulate the pressure and one flowrate, not necessarily the feed.

- Feed Rate and Pressure
- Vapour Rate and Pressure
- Liquid Rate and Pressure

This gives rise to six different alternatives. Each one is shown below with a short discussion on whether it is feasible or not. Note that the shaded valve indicates a valve used for inventory regulation. Not shown is the controller and level measurement. Since all measurements in this example orginate in the flash drum that end of the loop is omitted to simplify the diagrams.

Feed Rate and Pressure

Figure below below shows the alternatives for this set up.

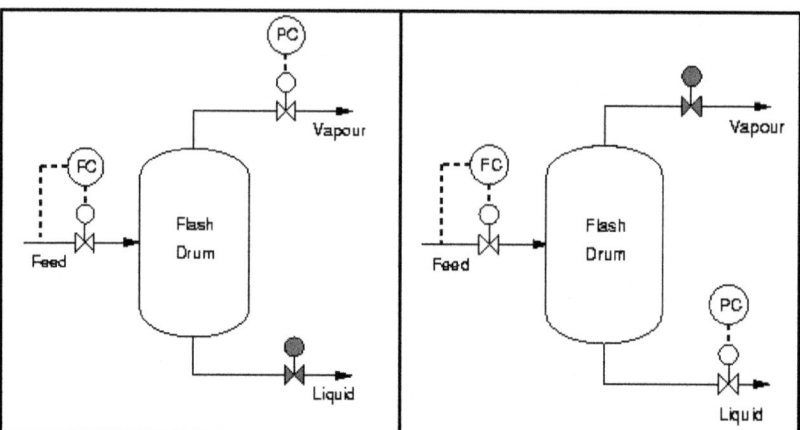

Fig. : Feed Rate Controlled.

Figure (a) is a very common arrangement and will work well. Both pressure and level loops have good adjustment-measurement sensitivity. There is some undesirable interaction because opening the pressure control valve increases the rate of boiloff and hence affects the level.

Figure (b) will not work. Although as noted above the vapour rate affects level, altering the liquid rate does *not* change the pressure of the flash.

Vapour Rate and Pressure

Figure below shows the alternatives for regulation of vapour rate and pressure.

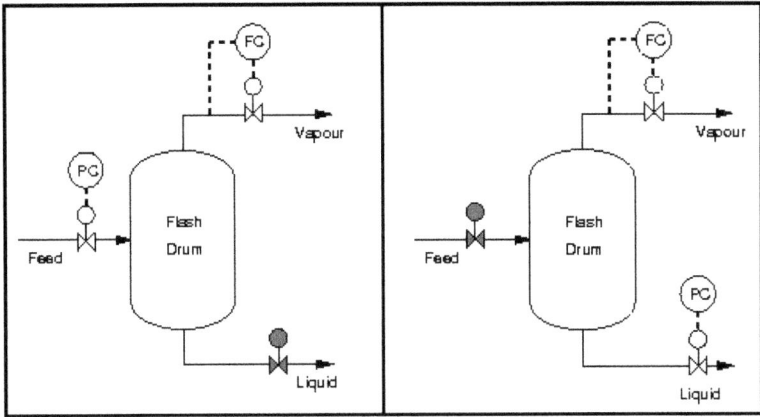

Fig. : Vapour Rate Controlled.

Figure (a) is another fairly conventional arrangement. Level control is good, although there is interaction between both flow and pressure loops. That is if the flow of vapour is increased then the pressure will decrease significantly and so the feed rate will have to be increased also. Decreasing the vapour flow will have the opposite effect causing the feed rate to be reduced.

Figure (b) is unworkable for the same reasons given above for figure (b).

Liquid Rate and Pressure

Finally figures gives the two alternatives for liquid rate and pressure.

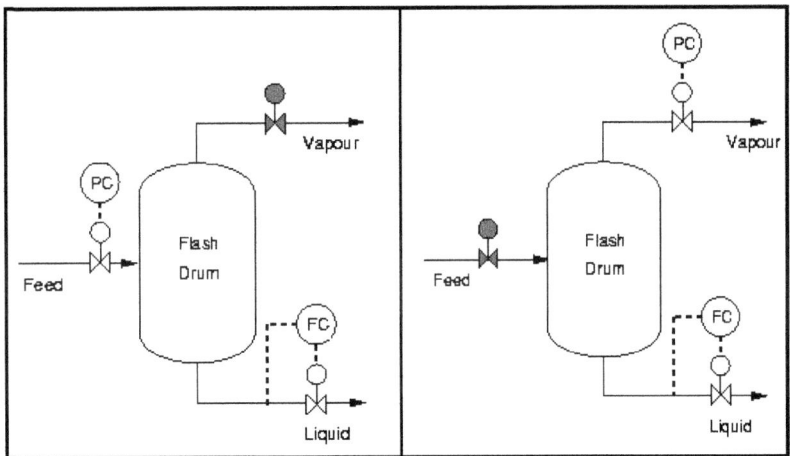

Fig. : Liquid Rate Controlled.

Both of these schemes will work. However figure (b) should give stronger response of both pressure and level control loops to their respective adjustments, and hence less interaction. It would be preferred to figure (a).

Alternative Arrangement

It would be possible to devise other schemes in which the flow control loop acted indirectly. For example, by adjusting a stream other than the one which is measured as shown below in figure. Such arrangements should be avoided, as should any system in which the operation of one loop, here the flow control, depends also on the operation of another, here the inventory loop.

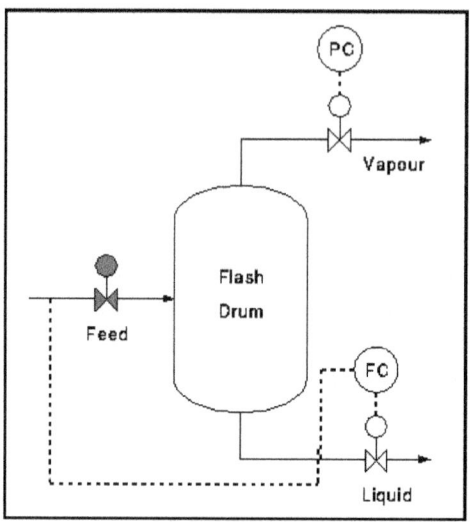

Fig. : Another Possibility?

Non-Adiabatic Flash: Degrees of Freedom

As with the Adiabatic Flash we will start by evaluating the control degrees of freedom from the equation

$$d.o.f = unknowns - equations$$

Unknowns

In this example there are 4 streams associated with the flash. There are 3 process streams each with unknown composition, temperature and pressure, and 1 energy flow. This gives

$$3 [C + 2] + 1 = 3C + 7 \; unknowns$$

Equations

In a flash with heat input there are

- C material balances
- C equilibrium relationships
- 1 energy balance

- 1 equation relating the temperature of the products *i.e.* they must be the same.
- 1 equation relating the pressure of the products *i.e.* they must also be the same.

Hence altogether there are

$$2C + 3 \; constraints$$

Degrees of Freedom

From the above we can evaluate that we have

$$C + 4 \; degrees \; of \; freedom$$

Now, as before, we can fix the composition, temperature and pressure of the feed stream. These C+2 constraints can be included in the above to give

$$3 \; control \; degrees \; of \; freedom$$

Once again this answer can be compared with that obtained directly.

$$C.D.F = no. \; of \; connections + 1 - no. \; of \; phases$$

- Number of connections for an flash with heat input = 4
- Number of phases = 2
- Hence C.D.F = 4 + 1 - 2 = 3, as expected!

Non-Adiabatic Flash: Control Strategy

The non-adiabatic flash separator has an additional adjustment not present in the adiabatic case, namely a heat input from a heating stream of steam or other utility. Analysis of the design problem, confirms that there is a further *degree of freedom* and therefore a flowrate plus two further variables can be set. *i.e.* there are four possible adjustments we can manipulate, the three mentioned and the material balance maintained with the fourth.

The best procedure to adopt in these examples is the following.

- Put in all flow controllers on the streams which they regulate.
- Check to see if it is now possible to locate the inventory control unambiguously, if so do it.
- Consider all possible combinations for remaining loops.
    - o Eliminate those without casual relationships between measurement and adjustment.
    - o Evaluate the others.

Note that while it is usually important to locate strategic loops before choosing the manipulated variable for level, we have found that liquid rate does not affect either temperature or pressure. All other things being equal, this would be the preferred choice for the level control loop. As with the adiabatic example, the measurement ends of the control loops are not shown and the adjustment for the level loop is indicated by a shaded valve.

Non-Adiabatic Flash: Examples

Below are four examples with the following parameters regulated.

- Pressure, temperature and vapour rate
- Pressure, liquid and vapour rates
- Pressure, liquid to vapour flow ratio and feed rate
- Pressure, temperature and vapour to feed ratio

Pressure, Temperature and Vapour Rate

Figure below shows the best arrangement for this option. In this case interchanging temperature and pressure loops could result in an inferior arrangement, as the feed rate would probably not have so direct an effect on temperature, but would still work.

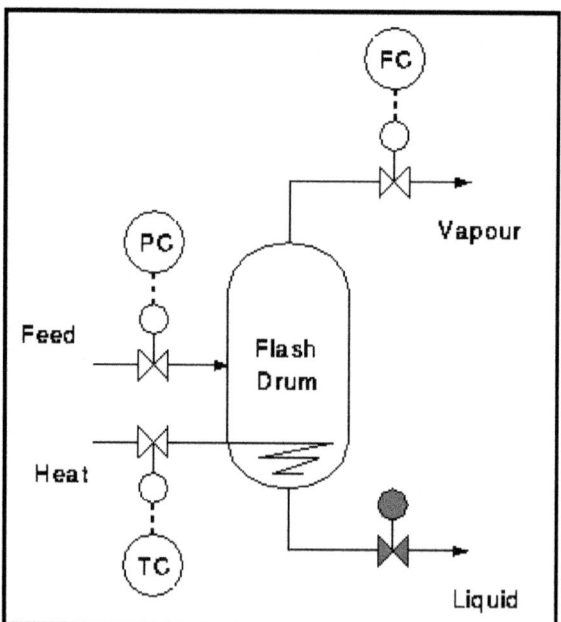

Fig. : Pressure, Temperature and Vapour Rate Regulated.

Pressure, Liquid and Vapour Rates

Once flow controllers are placed on the liquid and vapour product streams, the third material stream, *i.e.* the feed, *must* be used to maintain the mass balance. This leaves only the heat input, steam, valve to regulate pressure. This is the only possible scheme and is shown below in figure below.

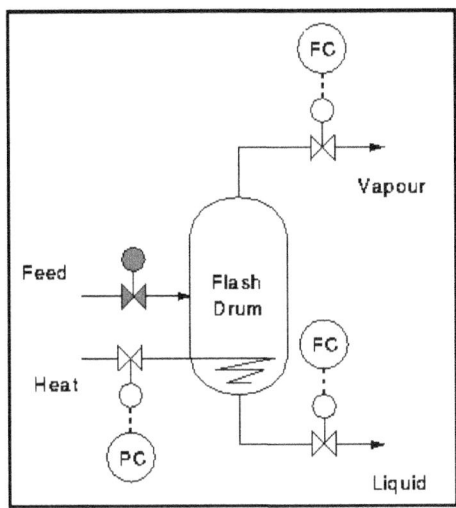

Fig. : Pressure, Liquid and Vapour Rates Regulated.

Pressure, Liquid to Vapour Flow Ratio and Feedrate

To regulate a ratio requires a flow control loop on one stream, with its setpoint adjusted by a ratio controller. A flow measurement on the second stream feeds into the ratio controller which *sees* both flows and changes the setpoint of the flow controller accordingly. Thus only one of the ratioed streams has a valve, but both have flowmeters. In principal the valve may be on either stream.

A suitable arrangement is shown in Figure below. The flow control valve for the ratio system has been located on the vapour line in order to leave the liquid line for level adjustment. Pressure control must then be by the heat input.

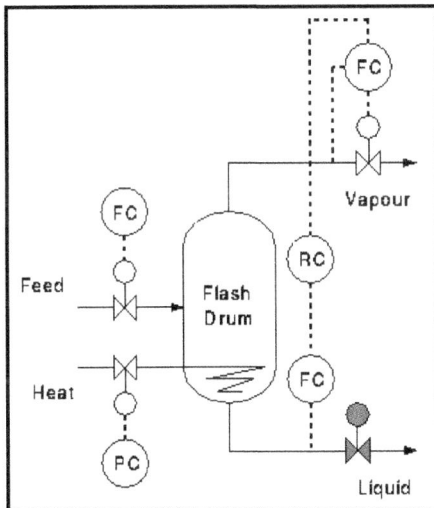

Fig. : Pressure, Liquid to Vapour Flow *Ratio* and Feedrate Regulated.

Pressure, Temperature and Vapour to Feed Rate Ratio

One alternative is shown in figure below. It would be possible here to put the valve for the ratio control on either feed or vapour streams. Also it is possible to interchange the temperature and pressure loops. The best choice would depend on the particular system: flowrates, component volatilities, *etc.* and would be determined after detailed modelling of the process.

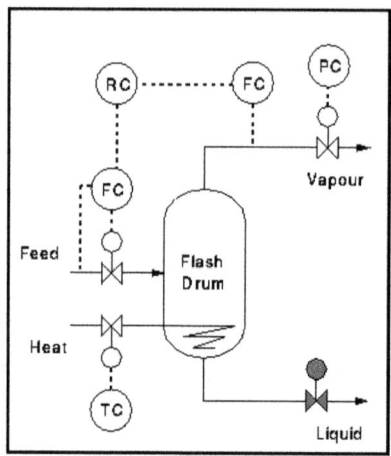

Fig. : Pressure, Temperature and Vapour to Feed Rate *Ratio* Regulated.

Three Phase Flash

In a three phase flash (or separator) the feed into the vessel separates into two liquid phases and a vapour phase. An example where this can be used is in the primary separation of light organics, heavy organics and water. The diagram below shows a typical three phase separator.

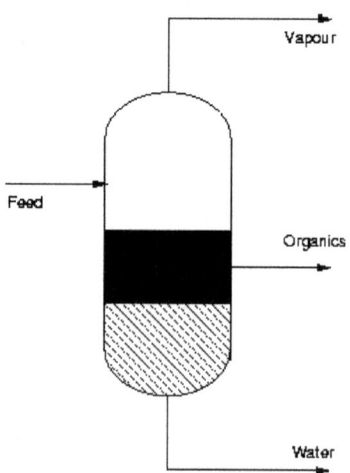

Fig. : Three Phase Flash.

Now let us consider the degrees of freedom:

C.D.F = no. of streams - no. of interfaces

- Number of streams = 4
- Number of interfaces = 2
- Hence C.D.F = 4 - 2 = 2

Typical control specifications would be feed rate and pressure. Note that **two** interfaces must be maintained with two control loops. An example of a control scheme is shown in the diagram below.

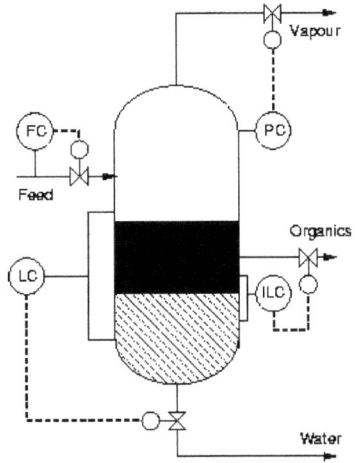

Fig. : Three Phase Flash With Control.

Countercurrent Cascade Columns

We have already looked briefly at countercurrent cascade columns in Module 2.1: Controlling Simple Processes. Here is a diagram of one without control.

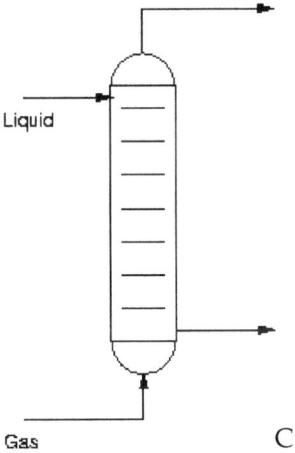

Fig. : Countercurrent Cascade Column.

Now let us consider control.

C.D.F = no. of streams - no. of interfaces

- Number of streams = 4
- Number of interfaces = 1 (gas-liquid)
- Hence C.D.F = 4 - 1 = 3

For gas-liquid operations the three strategic variables to control would be

- Column pressure
- One flow rate
- One composition

Below are three examples of controlling a cascade column with comments on how good or bad the control scheme is.

Example 1 - Good Composition Control

In this first example the control loops are:

- Liquid feed flowrate
- Column pressure - tops flowrate
- Bottoms composition - gas input
- Inventory/level - bottoms flowrate

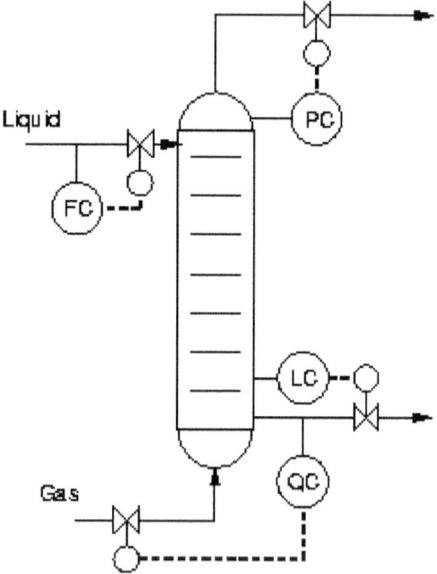

Fig. : Control Scheme 1.

This is a good control scheme. All the adjustments have a direct effect on the controlled variables. In particular look at the composition control of the bottoms product. As you can see the adjustment and measurement are both at the bottom of the column.

Example 2 - Poor Composition Control

In this second example the control loops are:

- Liquid feed flowrate
- Column pressure - tops flowrate
- Tops composition - gas input
- Inventory/level - bottoms flowrate

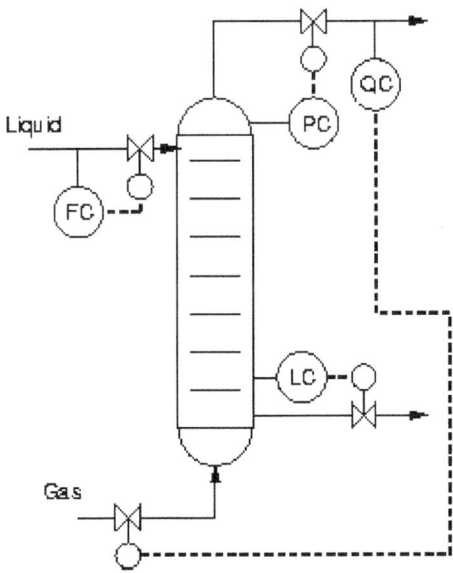

Fig. : Control Scheme 2.

This is not quite as good as the previous example. The pressure, flowrate and level are as before. However this time the top composition is controlled using the flowrate of the gas entering at the bottom of the column. Hence this time there will be a time delay between making the adjustment and the composition changing to reflect this change. This will be the length of time taken by the gas too travel up the column.

Example 3 - Bad Composition Control

Finally in this third example the control loops are:

- Gas feed flowrate
- Column pressure - tops flowrate
- Bottoms composition - liquid input
- Inventory/level - bottoms flowrate

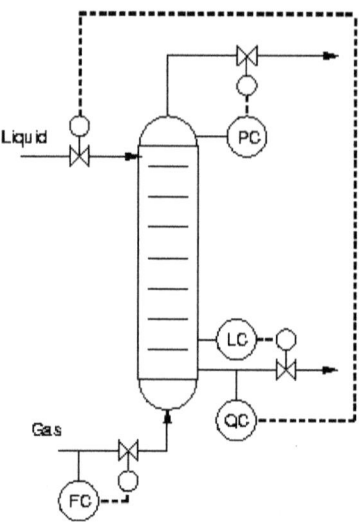

Fig. : Control Scheme 3.

In this example the flowrate of the gas is constant and once again the pressure and level control are as before. However this time the bottoms composition is regulated using the liquid flowrate entering at the top of the column. So once again there will be a time delay between making the adjustment and seeing the effect of the adjustment, but this time it will be the length of time taken for the liquid to travel down the column - a significantly longer time. Hence this control scheme should be avoided.

Liquid-Liquid Extraction Columns

Finally let us look at the situation where the gas in the previous example is replaced by another (lighter) liquid to form a liquid-liquid extraction column.

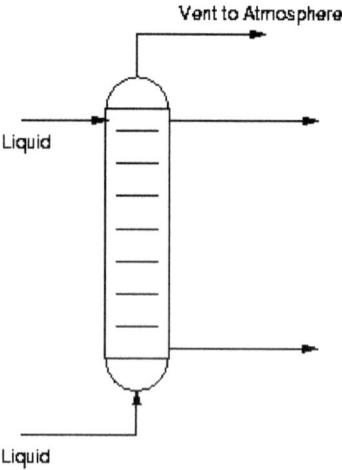

Fig. : Liquid-Liquid Extraction Column.

In this case note that there is going to be some vapour present at the top of the column. This gives another phase and so the degrees of freedom analysis gives us:

C.D.F = no. of streams - no. of interfaces

- Number of streams = 5
- Number of interfaces = 2 (vapour-liquid, liquid-liquid)
- Hence C.D.F = 5 - 2 = 3

A liquid-liquid column running full of both liquids would probably use *implicit* pressure regulation by leaving an open line to another part of the plant or alternatively may be vented to atmosphere. This effectively reduces the degrees of freedom to 2. Normally one flowrate and one composition of controlled along with the two interface control loops.

The example below shows the flowrate of the lighter liquid being controlled with the flowrate of the heavier liquid entering at the top of the column being used to regulate the composition of the top product. The two product streams are used for inventory.

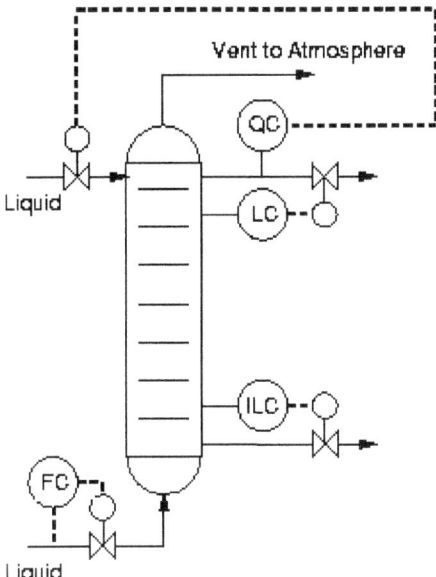

Fig. : Liquid-Liquid Extraction Column with Control.

Chapter 9

CONTROL SYSTEMS FOR COMPLEX PROCESSES

INTRODUCTION

The aim of this module is to introduce the control of distillation columns. We will start by analysing the degrees of freedom to establish how many and which control parameters it is possible to control and/or manipulate. Then we move on to discuss different ways to control the two most important parameters: composition at the top of the column and the pressure of the column. Finally there are a number of examples showing different control structures.

Degrees of Freedom Analysis

All we have to do is count all the streams in the process. Separately count the total number of extra phases *i.e.* add up all occurrences of phases greater than one in all units. The number of control degrees of freedom is the difference between these two numbers.

Figure below shows this method.

- Total Streams = 8
- Extra Phases = -3
- Degrees of Freedom = 5

So the number of degrees of freedom is 5. However, a typical control strategy for such a process would use only 4 of these - *feedrate, column pressure, top and bottom composition*. This is because the column and condenser are normally maintained at the same pressure.

However, a valve could be placed in the line between. This would actually be undesirable as reducing the condenser pressure will decrease the temperature driving force available from the cooling medium.

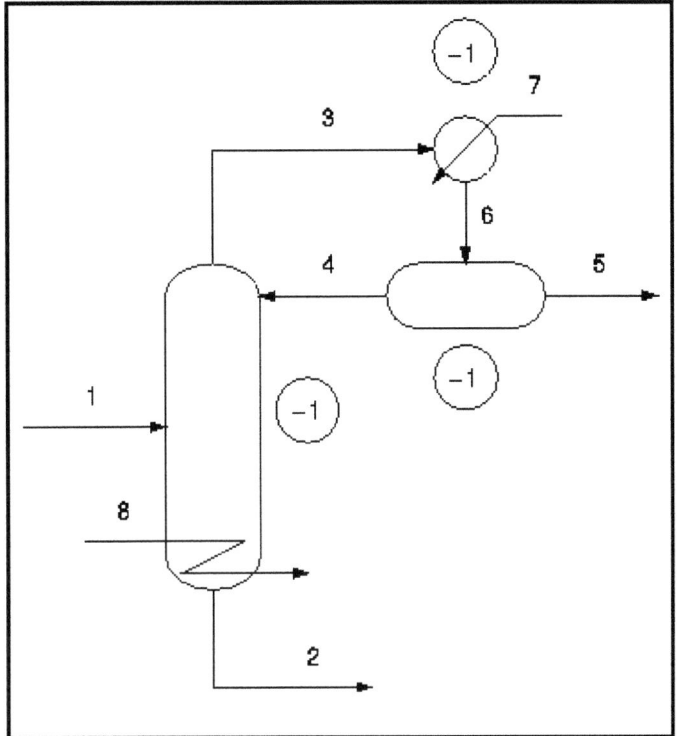

Fig. : Degrees of Freedom Analysis of Distillation Column.

CONTROLLING PRESSURE IN DISTILLATION

In a distillation column it is usually necessary to regulate the pressure in some way. Below there are five different methods described for doing this.

- Vent to Atmosphere
- Cooling Water
- Flooded Condenser - 1
- Flooded Condenser - 2
- Partial Condenser

One thing to note is that in none of them is a valve simply placed on the vapour line. This would lead to the use of a large expensive control valve. Instead the pressure is controlled indirectly involving the use of the condenser and/or reflux drum.

Vent to Atmosphere

Figure below shows the easiest way to control the pressure in a column operating at atmospheric pressure.

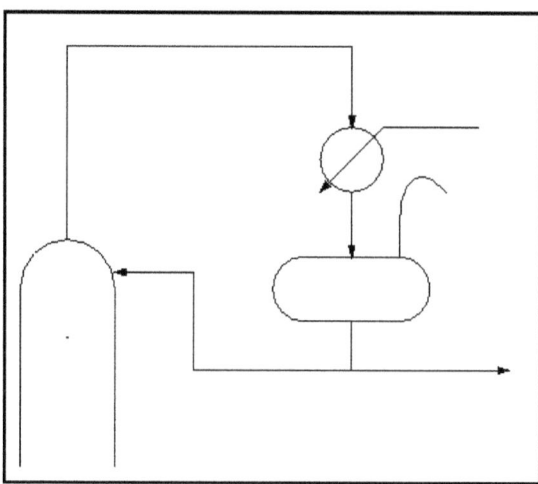

Fig. : Vent to Atmosphere.

In this case the cooling water flow stays constant and the reflux drum is vented to atmosphere. Thus the reflux drum and hence the top of the column are at atmospheric pressure. The advantage of this scheme is that it requires one less control valve. The disadvantage is that the tops have to be subcooled so that a minimal amount of vapour is lost through the vent. Hence more energy is required from the reboiler when the reflux is added to the top of the column.

Cooling Water

Figure shows the most common method for controlling the pressure - adjustment of the cooling water flow.

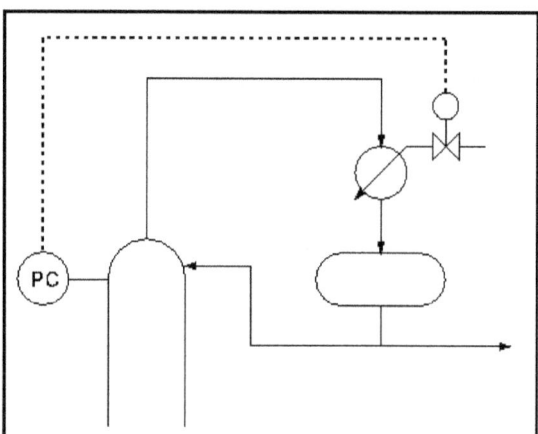

Fig. : Cooling Water.

In this case if the cooling water flow is increased then more vapour is condensed and the vapour pressure is reduced (and vice versa).

Flooded Condenser - 1

Figure below shows the classic flooded condenser approach.

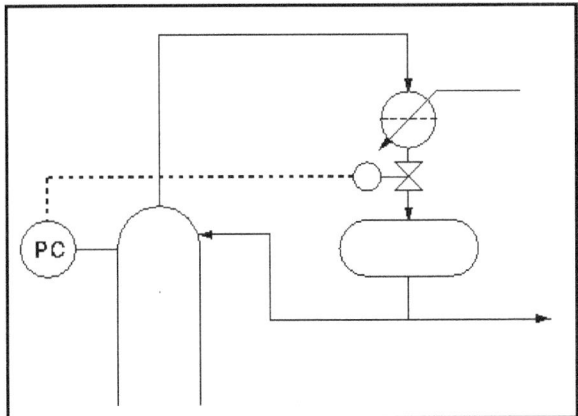

Fig. : Flooded Condenser - 1.

Again in this setup, as with the first example, there is no valve on the cooling water. Instead the valve is in the liquid line between the condenser and reflux drum.

If this valve is closed then the condensed vapour *i.e.* liquid will build up and *flood* the condenser. This has the effect of reducing the heat exchange area, thus reducing the amount of vapour being condensed and hence increasing the pressure.

The valve can then be opened, the liquid level will fall, increasing the heat exchange area and hence decreasing the pressure.

Flooded Condenser - 2

Figure below shows an alternative arrangement for a flooded condenser.

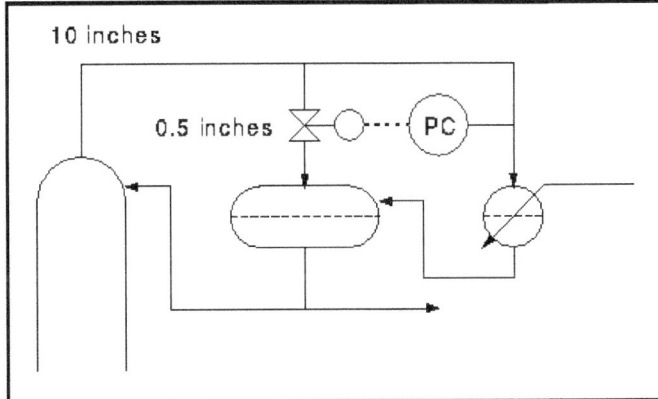

Fig. : Flooded Condenser 2.

The first thing to notice about this setup is that the reflux drum and condenser are at the same level. The second important point is that the vapour line, on which there is the control valve, is very small in comparison with the overhead line. If the valve is opened there is a small escape of gas into the reflux drum. This pushes the liquid level **down** in the drum and **up** in the condenser, flooding it and reducing the heat exchange area as in the last example.

Therefore to increase the pressure the valve is opened and to decrease the pressure the valve is closed.

Partial Condenser

The final example is the control of a partial condenser.

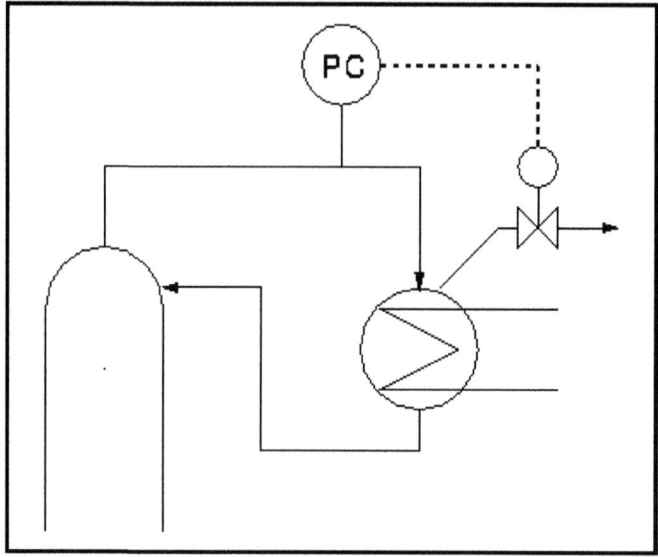

Fig. : Partial Condenser.

The above scheme is used if the overhead product is required as a vapour.

Controlling Tops Composition in Distillation

As well as pressure, the other parameter most likely to be controlled is the composition of the tops product. The reason is that the final product will most probably come from the top of the column and it is important to know its composition. Again, as with pressure, there are many different ways of controlling the tops composition. Three methods are described below.

- Reflux Rate
- Reflux Ratio
- Distillate Rate

Reflux Rate

In this first example the reflux rate is adjusted to control the composition of the tops product.

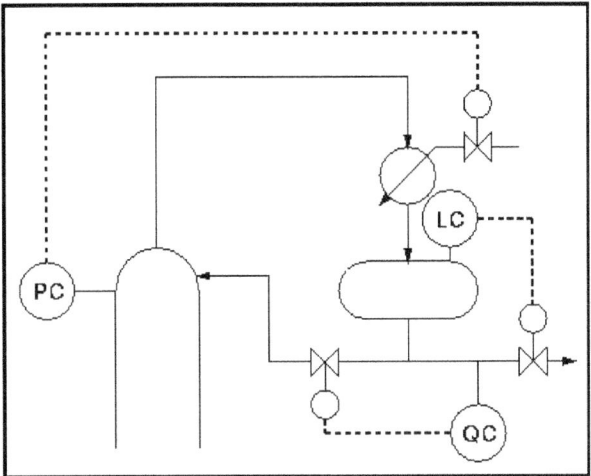

Fig. : Reflux Rate.

As the amount of reflux is changed so the temperature profile in the column changes and hence the composition.

Reflux Ratio

The second example uses the reflux ratio as the control parameter.

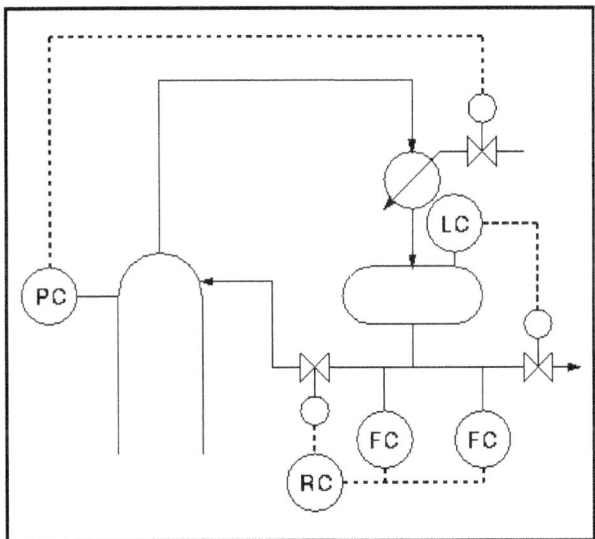

Fig. : Reflux Ratio.

When designing a distillation column it is usually the reflux ratio that is determined. This can be kept constant throughout operation by using two flow indicators and a ratio controller.

Distillate Rate

The third example is for high purity tops. It uses the distillate flowrate to control the distillate composition.

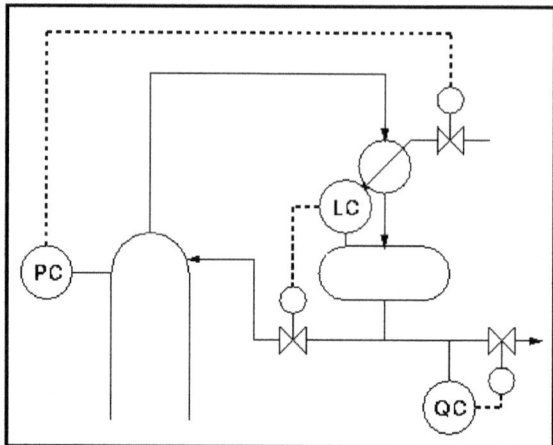

Fig. : Distillate Rate.

It can be shown that for a high purity column *i.e.* one with a large reflux, that the composition of the distillate is sensitive to the distillate flow but insensitive to the reflux rate. Therefore for a high purity column the control scheme outlined above is used. It should be noted that tight control on the level in the reflux drum is required using the reflux rate.

Distillation Column Control Examples

The following examples describe alternative control strategies of fairly standard form.

- Pressure, Overheads Rate and Composition
- Pressure, Bottoms Rate and Composition
- Pressure, Bottoms Rate and Overhead Composition, With Partial Condenser
- Pressure, Overhead Rate and Bottoms Composition
- Pressure, Bottoms Rate, Overhead Rate and Composition

In all cases actual composition controllers are shown. These could of course be replaced by inferential measurement from temperature, with or without cascade of a slower analyser. Unless otherwise stated, it has been assumed that the feed rate to the system is not available as a manipulated variable.

PRESSURE, OVERHEADS RATE AND COMPOSITION

This is a fairly standard configuration for a single product column, *i.e.* when the bottoms streams is a byproduct, recycle or goes to further processing.

Although the overheads composition is regulated by adjusting the steam rate at the base of the column, the response of the column to heat input changes is quite rapid, and so this strategy is acceptable.

Pressure control on condenser cooling water is shown; of course any other pressure control scheme would be acceptable.

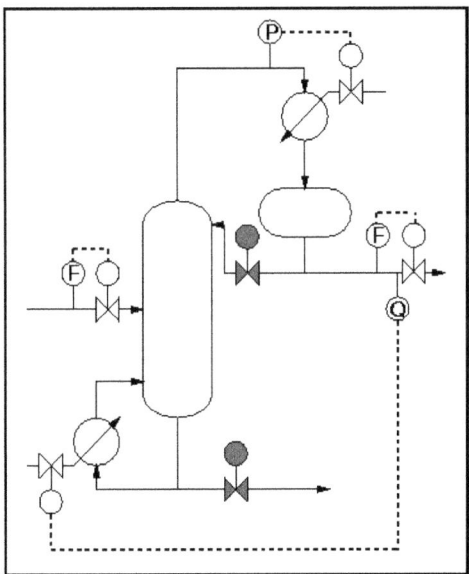

Fig. : Overheads Rate and Composition.

Pressure, Bottoms Rate and Composition

This is the analogous situation to the previous case, in the rather less usual circumstances where a main product is withdrawn from the bottom of the column.

This does not work well, since either the bottom level, as here, or composition, has to be regulated by adjusting the reflux rate. In either case the loop involves a long delay due to the hydraulic lags on each tray.

It is probably marginally better to regulate composition by steam rate since this is a more important quantity than level, although the two loops could be interchanged with the steam adjusting the level, which is quite a good scheme, and the reflux manipulating the bottoms composition, which is very poor. Fortunately this is an unusual requirement, as main products normally come from the top of columns for other reasons.

A standard flooded condenser pressure control system is shown.

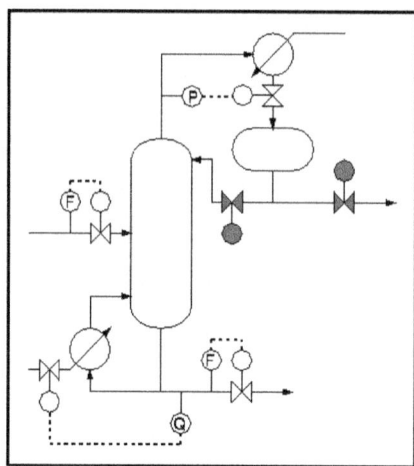

Fig. : Bottoms Rate and Composition.

Pressure, Bottoms Rate and Overhead Composition, With Partial Condenser

This is not a particularly common strategy, but the arrangements for a column with partial condenser are typical. The pressure in such a system is almost always manipulated by a valve on the vapour product line. There is no reflux drum, and reflux rate is often set implicitly by adjusting the cooling load on the condenser.

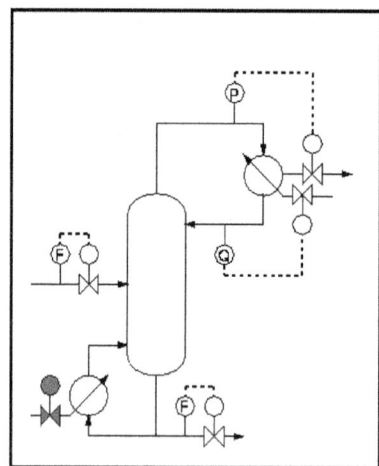

Fig. : Bottoms Rate and Overhead Composition, With Partial Condenser.

Pressure, Overhead Rate and Bottoms Composition

This scheme should work satisfactorily as all adjustments are made at the same end of the column as the related measurements. The pressure control scheme is the so-called *hot gas bypass*. Note that the layout of condenser and reflux drum shown is critical to the operation of this method, which is actually a variation on the *flooded condenser* approach. The bypass is a very small pipe which bleeds vapour into the reflux drum where it does *not* immediately condense. The pressure in the system *rises* as the bypass valve is opened.

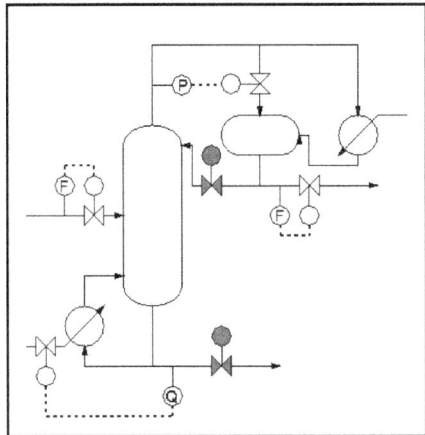

Fig. : Overhead Rate and Bottoms Composition.

Pressure, Bottoms Rate, Overhead Rate and Composition

Since *three* regulated quantities are specified, the feed to the unit must be available as an adjustment. Apart from this, the arrangements are similar to those of the first example. Level control on the column base is not very satisfactory due to the lags between the feed and the bottom of the column, but any other arrangement would be worse.

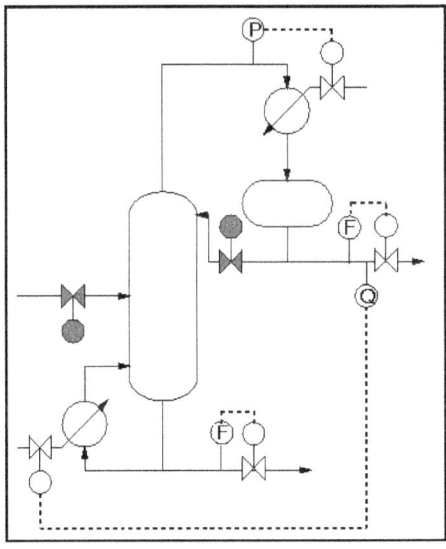

Fig. : Bottoms Rate, Overhead Rate and Composition.

EXTENSION OF HIERARCHICAL DECOMPOSITION

In this module we aim to introduce an **alternative** method for the design of control systems for complex processes. In the **traditional** approach the process engineer is presented with a complete flowsheet and is required to specify the structure of all the loops.

There are two disadvantages to this method.

- The complexity of the task.
    - o It is very difficult to choose the control structures for a complete flowsheet from amongst the large number of alternatives.
    - o It is made even more difficult in that there is no defined procedure for tackling this problem.
- It is impossible for any feedback from the control study to be taken into account.
    - o Decisions about equipment will already have been taken.
    - o There will be considerable reluctance to change the process flowsheet to accomodate possible control problems.

For these reasons a **Hierarchical** approach to the design of control systems will be used. In this method controls are placed, as far as possible, on the flowsheet in its simplest form. As the flowsheet is elaborated more loops are added.

This approach has significant advantages over existing practice.

- The task of placing all control loops on a complete and detailed flowsheet is reduced to more manageable proportions.
- Attention is concentrated on strategically important control loops rather than those of secondary importance such as level in holding tanks.
- Decisions on process and control system structures are taken together, allowing the latter to influence the former in a manner not possible with the conventional approach.

Outline of Procedure

This hierarchical approach is already well established in the conceptual design of chemical flowsheets. In the Douglas scheme the designer starts with a single block showing only feeds and products and proceeds by sequential elaboration through the following steps:

- Input-Output structure
- Recycle structure
- Separation sequencing
- Energy integration

The method used for designing control systems employs a similar set of steps:

- Feed and product rate control
- Recycle rates and composition
- Product and intermediate stream composition
- Temperature and energy balance control
- Inventory regulation

As with the design outline above, these controls are placed on the flowsheet at different stages in the structure.

- Input-Output Structure
- Recycle Structure
- Functional Subsystems Structure

Advantages of this hierarchical approach are:

1. At any one time only a subset of the overall problem is considered. This simplifies the problem somewhat and avoids the *intellectual overload* inherent in the conventional approach.
2. The more important **strategic** control loops are emphasised, *i.e.* those affecting the economic performance of the plant such as product rate, overall conversion and product quality.
3. Less important decisions are left to the end, *i.e.* those associated with operability such as regulation of inventory.
4. By proceeding in parallel with both process and control system design, alternatives for both may be considered at each step.

We will now go through these steps in more depth and discuss what should be considered at each level.

Input-Output Structure

At this stage we can only look at control of the feed and product. However this affects the basic shape of the process and so control decisions at this stage determine the whole strategy. Things to think about at this point are:

- Identify likely disturbances
  - o magnitude
  - o timescale
- Ranges and Scaling
- Sensitivities *i.e.* Steady-state gains

Recycle Structure

The next stage in the design of the control system is to elaborate it further to include the recycle structure. In some cases this may be done in more than one step. Things to consider at this stage are:

- Sensitivities and ranges
- Interactions *i.e.* RGA etc
- Functional controllability
- Multiple forward paths and possible inverse response
- Approximate dynamics
  - o Time delays
  - o Approximate time constants
  - o Simple models

Functional Subsystems Structure

The next stage is to include sections such as feedsystems, reactors or separation schemes. Important points which can be added to the lists above are:

- Simple dynamic unit models
- Standard control schemes investigated for subsystems

This last point covers systems such as feed/mixing systems or separation schemes. It includes equipment such as flash units or distillation columns.

Discussion

At each of the above three stages it is important to understand what each stream is there for and to ensure that as much control as possible has been included. Thus the following steps are used:

- Consider each stream on the flowsheet at the current level.
- Why is it there?
- What will decide its flowrate?
- Can a complete, partial or tentative control system be added to manipulate its flow and meet the above objective?
- Draw in as much of the control system as possible.

This is an extremely straightforward procedure. It is easy to see what all streams do when the flowsheet is in a simplified form. On a complete PFS or ELD it will be much harder to identify important streams and their interrelationships. Similarly, the purpose of streams is most clearly established in the designer's mind at the point when he has just identified them in the process design.

The final stages in the design of a control system are to consider energy integration and inventory control.

For energy integration the general principal is to derive a control system for an integrated energy recovery network from the unintegrated form by replacing valves on utility streams with bypasses. The case of energy integrated distillation can be rather more interesting. For example, in unintegrated distillation systems reboiler heat load is often used as a manipulated variable for bottom product composition regulation. Condenser heat load is used to regulate column pressure. However, what happens when the condenser of the first column forms the reboiler of the second?

We regard the regulation of inventory, *i.e.* ensuring that all mass balances do in fact balance, as the lowest level of control, to be achieved after all *strategic* control systems have been specified. It is not usually difficult to place level control loops once the number of alternatives has been reduced.

Consider a process or subsection with n inputs and outputs which contribute to a single total material balance. $(n-1)$ of these may be set by strategic control loops for composition, temperature *etc.* One remaining stream must be used to ensure material balance *via* level control in a liquid system or pressure control in a gas system or implicitly by some inherent control mechanism such as the weir on a distillation tray.

Chapter 10

THE COMPONENTS OF A CONTROL LOOP

COMPONENTS OF A CONTROL LOOP

A controller seeks to maintain the measured process variable (PV) at set point (SP) in spite of unmeasured disturbances (D). The major components of a control system include a sensor, a controller and a final control element. To design and implement a controller, we must:

1) have identified a process variable we seek to regulate, be able to measure it (or something directly related to it) with a sensor, and be able to transmit that measurement as an electrical signal back to our controller, and

2) have a final control element (FCE) that can receive the controller output (CO) signal, react in some fashion to impact the process (*e.g.*, a valve moves), and as a result cause the process variable to respond in a consistent and predictable fashion.

Home Temperature Control

The home heating control system can be organized as a traditional control loop block diagram. Block diagrams help us visualize the components of a loop and see how the pieces are connected.

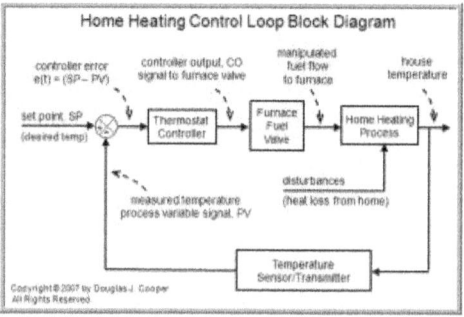

A home heating system is simple on/off control with many of the components contained in a small box mounted on our wall. Nevertheless, we introduce the idea of control loop diagrams by presenting a home heating system in the same way we would a more sophisticated commercial control application.

Starting from the far right in the diagram above, our process variable of interest is house temperature. A sensor, such as a thermistor in a modern digital thermostat, measures temperature and transmits a signal to the controller.

The measured temperature PV signal is subtracted from set point to compute controller error,

$$e(t) = SP - PV.$$

The action of the controller is based on this error, $e(t)$.

In our home heating system, the controller output (CO) signal is limited to open/close for the fuel flow solenoid valve (our FCE). So in this example, if $e(t) = SP - PV > 0$, the controller signals to open the valve. If $e(t)) = SP - PV < 0$, it signals to close the valve. As an aside, note that there also must be a safety interlock to ensure that the furnace burner switches on and off as the fuel flow valve opens and closes.

As the energy output of the furnace rises or falls, the temperature of our house increases or decreases and a feedback loop is complete. The important elements of a home heating control system can be organized like any commercial application:

- Control Objective: maintain house temperature at SP in spite of disturbances
- Process Variable: house temperature
- Measurement Sensor: thermistor; or bimetallic strip coil on analog models
- Measured Process Variable (PV) Signal: signal transmitted from the thermistor
- Set Point (SP): desired house temperature
- Controller Output (CO): signal to fuel valve actuator and furnace burner
- Final Control Element (FCE): solenoid valve for fuel flow to furnace
- Manipulated Variable: fuel flow rate to furnace
- Disturbances (D): heat loss from doors, walls and windows; changing outdoor temperature; sunrise and sunset; rain…

A General Control Loop and Intermediate Value Control

The home heating control loop above can be generalized into a block diagram pertinent to all feedback control loops as shown below.

Both diagrams above show a closed loop system based on negative feedback. That is, the controller takes actions that counteract or oppose any drift in the measured PV signal from set point.

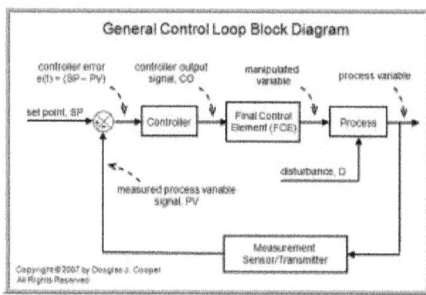

While the home heating system is on/off, our focus going forward shifts to intermediate value control loops. An intermediate value controller can generate a full range of CO signals anywhere between full on/off or open/closed. The PI algorithm and PID algorithm are examples of popular intermediate value controllers.

To implement intermediate value control, we require a sensor that can measure a full range of our process variable, and a final control element that can receive and assume a full range of intermediate positions between full on/off or open/closed. This might include, for example, a process valve, variable speed pump or compressor, or heating or cooling element.

Note from the loop diagram that the process variable becomes our official PV only after it has been measured by a sensor and transmitted as an electrical signal to the controller. In industrial applications. these are most often implemented as 4-20 milliamps signals, though commercial instruments are available that have been calibrated in a host of amperage and voltage units.

With the loop closed as shown in the diagrams, we are said to be in automatic mode and the controller is making all adjustments to the FCE. If we were to open the loop and switch to manual mode, then we would be able to issue CO commands through buttons or a keyboard directly to the FCE. Hence:

- open loop = manual mode
- closed loop = automatic mode

Cruise Control and Measuring Our PV

Cruise control in a car is a reasonably common intermediate value control system. For those who are unfamiliar with cruise control, here is how it works.

We first enable the control system with a button on the car instrument panel. Once on the open road and at our desired cruising speed, we press a second button that switches the controller from manual mode (where car speed is adjusted by our foot) to automatic mode (where car speed is adjusted by the controller).

The speed of the car at the moment we close the loop and switch from manual to automatic becomes the set point. The controller then continually computes and transmits corrective actions to the gas pedal (throttle) to maintain measured speed at set point.

It is often cheaper and easier to measure and control a variable directly related to the process variable of interest. This idea is central to control system design and maintenance. And this is why the loop diagrams above distinguish between our "process variable" and our "measured PV signal."

Cruise control serves to illustrate this idea. Actual car speed is challenging to measure. But transmission rotational speed can be measured reliably and inexpensively. The transmission connects the engine to the wheels, so as it spins faster or slower, the car speed directly increases or decreases.

Thus, we attach a small magnet to the rotating output shaft of the car transmission and a magnetic field detector (loops of wire and a simple circuit) to the body of the car above the magnet. With each rotation, the magnet passes by the detector and the event is registered by the circuitry as a "click." As the drive shaft spins faster or slower, the click rate and car speed increase or decrease proportionally.

So a cruise control system really adjusts fuel flow rate to maintain click rate at the set point value. With this knowledge, we can organize cruise control into the essential design elements:

- Control Objective: maintain car speed at SP in spite of disturbances
- Process Variable: car speed
- Measurement Sensor: magnet and coil to clock drive shaft rotation
- Measured Process Variable (PV) Signal: "click rate" signal from the magnet and coil
- Set Point (SP): desired car speed, recast in the controller as a desired click rate
- Controller Output (CO): signal to actuator that adjusts gas pedal (throttle)
- Final Control Element (FCE): gas pedal position
- Manipulated Variable: fuel flow rate
- Disturbances (D): hills, wind, curves, passing trucks…

The traditional block diagram for cruise control is thus:

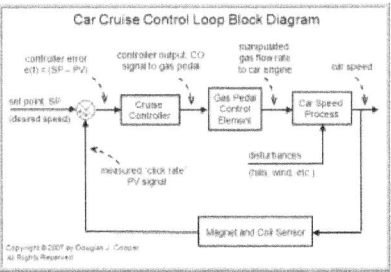

Instruments Should be Fast, Cheap and Easy

The above magnet and coil "click rate = car speed" example introduces the idea that when purchasing an instrument for process control, there are wider considerations that can make a loop faster, easier and cheaper to implement and maintain. Here is a "best practice" checklist to use when considering an instrument purchase:

- Low cost
- Easy to install and wire
- Compatible with existing instrument interface
- Low maintenance
- Rugged and robust
- Reliable and long lasting
- Sufficiently accurate and precise
- Fast to respond (small time constant and dead time)
- Consistent with similar instrumentation already in the plant

FINAL CONTROL ELEMENT

A final control element is defined as a mechanical device that physically changes a process in response to a change in control system setpoint. Final control elements relevant to actuators include valves, dampers, fluid couplings, gates, and burner tilts to name a few. Final control elements are an essential part of process control systems, allowing an operator to achieve a desired process variable output by manipulating a process variable setpoint.

Current trends in industry focus on improving quality and efficiency in an effort to reduce process costs. Many industrial facilities are recognizing the bottom line savings found from improved process control performance through precise and consistent control of final control elements.

Traditional methods of positioning final control elements include electric actuators with squirrel-cage motors that suffer from duty-cycle limitations, poor resolution, and poor reliability. Commonly used pneumatic actuators suffer from stick/slip overshoot and high maintenance requirements. While it may seem like a given that your actuator should move when and where the controller tells it to, in many cases dead time and stick/slip overshoot cause delays in reaching the setpoint. The inefficiencies of these technologies can subject the process to relatively poor control and unplanned outages causing process variability and increased process costs. Many times these costs go unnoticed because they are not direct and immediately apparent. When selecting an actuator for a final control element, the actuator should have performance characteristics that will enable a control system to perform as designed.

The key final control element actuator performance characteristics are as follows:

- Precise, repeatable positioning typically better than 0.15% of span.
- The ability to start and stop instantaneously without dead time or position overshoot.

- Continuous duty rating without limitations on the number of starts per minute.
- Perform consistently and unaffected by load.
- Rugged industrial design capable of operating in difficult environments without an effect on performance.
- Minimal periodic maintenance required.

A final control element actuator designed with these characteristics provides two extremely important advantages:

1. An ability to follow the demand signal from the controller precisely and instantly. This ensures that the actuator responds exactly as directed by the controller. Thus, the actuator is not the limiting factor in the control loop and the controller can function to its optimal levels.

2. A high degree of maintenance-free reliability. An actuator designed to function as outlined above by default is more rugged than typical actuators. By design, then, it is capable of a much higher degree of reliability.

Instrumentation technologists and mechanics

Instrumentation technologists, technicians and mechanics specialize in troubleshooting and repairing and maintenance of instruments and instrumentation systems.

TURN PROBLEM LOOPS INTO PERFORMING LOOPS

Each control system control loop contains five pieces: a sensing element, transmitter, controller, final control element, and process. Only when all five elements are performing their best will the control system meet expectations.

Frequently process control requires controlling one of four variables; flow, pressure, temperature, or level.

Flow measurement devices available today are more forgiving in their installation requirements than devices available even five years ago; so it may be that replacing a flow measurement device is justified on its ability to perform in the current installation.

When auditing existing flow measurement installations, information necessary to understand performance expectations includes:

- Full range (0-100%) is the minimum and maximum flow that passes by the measurement point;
- Rangeability (turndown) is the ratio between the maximum and minimum control points;
- Repeatability, frequently described in percent, is the ability to achieve the same output when the same input, coming from the same direction is provided; and

- Accuracy, often stated in either percent of full scale or percent of reading.

If, for example, flow was originally to be controlled at 1,400 gpm (gallons per minute) (5,300 lpm (liters per minute)) at

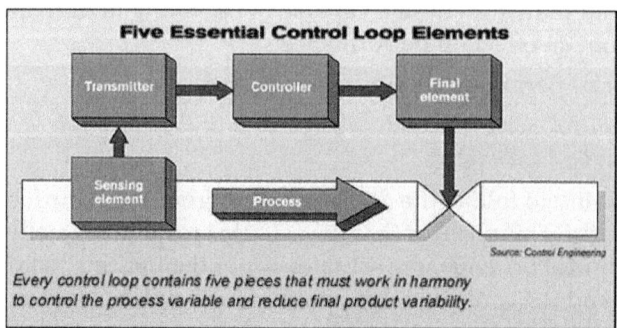

Five Essential Control Loop Elements

Every control loop contains five pieces that must work in harmony to control the process variable and reduce final product variability.

Flow measurement devices can be segregated into two categories: devices measuring flow rate and those measuring velocity.

Devices measuring flow are often head-type devices that depend on measuring pressure differences across inline interferences (*i.e.*, orifice plate) and typically provide about 3:1 rangeability. Accurate flow measurements in head-type devices occurs as long as the pressure and/or temperature of the flowing media remains at design (base or standard) conditions.

Velocity based measurement devices frequently provide measurements in mass, mole, or volume terms; and include magnetic-, vortex-, mass-, and turbine-meters. Devices in the velocity category rely on the basic formula *Mass = volume x density* .

Mass-flow measurement relies on the laws of nature that prohibit a stream from accumulating or loosing mass; thus mass-flow measurements are independent of changes in temperature, pressure, or pipe size. Units used in mass-flow measurement are usually pounds or kilograms with time periods of seconds, minutes, or hours such as pounds per hour (pph) or kilograms per second (kg/s).

When volume measurements are made assuming design conditions, an inaccurate measurement is produced. Overcoming these inaccuracies requires on-line compensation for density changes. For liquids, that means temperature compensation; and for vapors or gases, that means pressure and temperature compensation. Volume measurements are usually in cubic feet, cubic meters, cubic liters, or gallons with time periods of seconds, minutes, or hours, *i.e.*, cfh (cubic feet per hour), m3/s (cubic meters per second), or gpm (gallons per minute).

Mole-flow measurements are determined by the formula *Mole flow rate = mass flow rate ÷ molecular weight of the fluid* . Typical units for mole measurements are moles per second (mol/s) and kilomoles per hour (kmol/h).

Rangeability of velocity based flowmeters is improving, but a few years ago the rules-of-thumb were: magnetic meters (30:1), vortex devices (15:1), massflow meters (100:1), and turbine meters (10:1).

Flow Installation Rules of Thumb

Flow meter type	Straight pipe diameter requirements Up-stream	Straight pipe diameter requirements Down-stream
Note: Pipe diameter must match flowmeter size. Source: Control Engineering		
Magnetic meter	5	2
Mass flow meter	1	1
Vortex meter	10	25
Turbine meter	15	10
Orifice plate with 0.5 orifice to pipe diameter ratio.	25	4
Orifice plate with 0.7 orifice to pipe diameter ratio	40	5

Streamline or Turbulent Flow

Osborne Reynolds (1842-1912) determined turbulence influenced obtaining repeatable flow measurements. Reynolds developed a formula that when basic units are assigned to each quantity, the ratio is a dimensionless number. Reynolds determined that turbulence essentially disappeared and flow became streamline (laminar) at about 2,000. Between 2,000 and 4,000 measurement performance is questionable. Above 4,000, turbulence is good.

Producing turbulent flow is not enough for accurate flow measurements; how the turbulence is developed also influences the measurement. Valves, pumps, or piping configurations located close to flow sensors can cause unwanted flow stream influences.

Determining if a specific flowmeter installation can provide the required capability is best determined by referring to specific device installation instructions. When instructions are unavailable, 'rules-of-thumb' may help decide if sufficient up- and down-stream runs of pipe the same size as the flowmeter are installed.

Example Control Valve Flow Calculations

	Minimum	Normal	Maximum	Rangeability
Assumes a minimum control point of 375 GPM, a normal control of 750 GPM, and a maximum of 1,500 GPM. Source: Control Engineering				
Quick opening	10% open	25% open	50% open	5:1
Linear	10% open	50% open	80% open	8:1
Equal percentage	5% open	70% open	90% open	18:1

Error Sources

Primary sensors are frequently connected to transmitters that are used as transducers, receiving information in one form and converting it to another form.

For example, an RTD (resistance thermal detector) primary sensor provides temperature measurements in ohms/deg. Connected to a transmitter (transducer) the ohms/deg value is converted to 4-20 mA and transmitted to an indicator, recorder or controller.

A source of error in control loops occurs when transmitter calibrations are made at the electronics, and do not include the sensor. For example, it is common to find older thermocouples have drifted several degrees. Substituting a calibrating source for the thermocouple input will not reveal an inaccurate thermocouple.

Sensor interchangeability is another source of temperature measurement error. Standard temperature sensors allow for a reasonable tolerance around an 'ideal' sensor curve. Matched temperature sensors cost more but deliver significantly better accuracy. Be aware that some manufacturers deliver high accuracy systems by matching the sensor and transmitter to form a system. Extra care is required in maintaining systems using matched sensors to preserve the 'paid for' capability.

Primary sensors are the number one influence on control-loop performance-but final control elements rank a close second.

Final Control Elements

Final control elements come in a variety of shapes and sizes including variable-speed drives, heaters, and valves. Valves include globe, characterized ball, quarter-turn, butterfly, eccentric-disk, and knife gate.

Final control elements can be segregated into three performance classes; linear, equal percentage, and quick opening. Linear elements include globe and eccentric-disk valves, variable-speed drives, and heaters. Equal-percentage elements include globe, characterized ball, and butterfly valves. Included in the quick-opening class are globe, quarter-turn ball, knife gates, and dampers.

Globe valves appear in all three classes because they can be fitted with a variety of plug, seat, and cage designs to meet a broad range of applications; a point worth remembering as pro-cess audits are conducted and the need for changes analyzed.

Control valves have long been the primary final element installed to control flow, temperature, pressure, and level. Experience reveals that about 60% of installed control loops reach 100% of the measured variable range with only 30% of the final control element travel. That means a lot of businesses have bought more control valve capacity than necessary. When sizing and selecting control valves, it is best to calculate minimum, maximum, and normal flow for all three characteristics. With results side-by-side, the characteristic that provides the most uniform process gain is easier to determine. A valve that's too small will not pass the required flow, while one that is too large may result in unstable performance as it tries to control at very low increments of travel.

Since the goal is to reduce process variability, ensuring smooth control valve performance is critical to success.

Making permanent improvements to control-loop performance requires verifying that data are repeatable. Processes unable to repeat data often indicate problems in the measurement system and/or control valves. Two common causes of nonlinear response in control valves are excess hysteresis and stick-slip.

Hysteresis is the inability of a device to return to a previously established position when the input to the device is repeated. In control valves, hysteresis is distinguished from deadband by expecting that small reversals of input may not produce reversals of valve travel. Integrating processes often demonstrate an oscillatory behavior caused by control valves with excess hysteresis; self-regulating processes rarely do.

Sources to check for excess hysteresis include:

- Valve packing-gland tightness;
- Seal friction in rotary valves;
- Worn or loose linkages and couplings;
- Defective I/P (current-to-pneumatic) transducer;
- Inadequate supply air pressure;
- Defective positioner;
- Incorrect valve actuator bench set; and
- Undersized actuator.

Stick-slip cycling occurs when controller integral action continuously increases the controller output without a corresponding change in the actual valve position (stick phase). When the valve finally moves, it 'pops' and the process variable overshoots the setpoint (slip phase). Controller integral action drives the output in the other direction, setting up a distinctive continuous oscillation.

Controller Influences

Controllers exist to maintain a measured variable equal to a setpoint. The affects filtering have on controller performance is a seldom addressed topic. For example, transmitters often provide means of 'snubbing' the measured variable. Snubbing is accomplished using an adjustable orifice (or partially closed isolation valve) in the sensing line to reduce process pulsations from reaching the sensing element.

Some transmitters provide electronic filtering of the output signal. Many digital control systems allow users to apply one or more filters on input signals.

Regardless of the form, filters add lag to the signal and, when inappropriately used, can mask measurement variability and create unsafe conditions.

If examination of the raw measured-variable input indicates frequent, random spikes which cannot be eliminated at their source, then application of a filter as near the source of the noise is appropriate. The amount of filter should be the minimum necessary to filter excessive, but not all the noise from the signal. Use extra care when applying filters to integrating processes. The additional lag introduced by the filter can be canceled by the controller's derivative action.

Temperature measurements seldom contain excessive noise because of process measurement lags. If high-frequency noise is discovered on temperature measurement signals the cause is likely improper shielding of thermocouple leads. Source of the noise should be fixed, rather than applying filters. When filters must be applied to temperature measurement signals, they should be very small values.

Controller scan periods can be a source of poor control loop performance. As a rule-of-thumb, controller scan periods should be at least eight times faster than the loop time constant.

Conducting systematic audits of the five parts of existing control loops can pay big dividends in understanding the 'knobs' available to operations and the influence each contributes to process variability.

COMMON CAUSE OF CONTROL LOOP CYCLING

As many as one in five control loops demonstrates a continuous cycling at steady state when tuned with the optimum PI or PID tuning parameters calculated using any of the popular methods including Lamda and Ziegler-Nichols. In most cases, the cycling can be directly traced to nonlinear behavior of pneumatically actuated control valves. The two most common types of nonlinear control valve responses are hysteresis with deadband and stick-slip.

Hysteresis with deadband will cause steady-state cycling in properly tuned integrating loops, while stick-slip causes the same in self-regulating loops.

Stick-slip response is common in pneumatically actuated control valves using pneumatic positioners.

By design, pneumatic positioners are nonlinear devices. When a constant ramp input signal is applied to a pneumatic positoner, the gain is small and loads the actuator slowly. When the ramp input exceeds a predetermined value, the gain increases and loads the actuator dome at a faster rate.

Stick-slip occurs when the controller integral action continuously increases the controller output without a corresponding change in the actual valve position. When the valve finally moves, it pops and the process variable overshoots the setpoint. The error becomes negative and the controller integral action drives the output in the other direction. This results in the distinctive continuous limit cycle known as a stick-slip cycle. The process variable appears as a square wave oscillating around the setpoint. The controller output appears as a triangular wave with a frequency dependent on the tuning parameters, the valve, and the process gain.

There are three traditional solutions to stick-slip cycling problems.

Repair or replace the valve;

Place the controller in manual; or

Detune the controller integral setting.

Detuning the integral setting eliminates stick-slip cycling but also slows the control loop's ability to respond to setpoint changes.

Techmation Inc. (Scottsdale, Ariz.) has developed a deadband reset scheduling (DSR) algorithm that adjust controller integral settings depending on the size of the error between the setpoint and process variable.

MORE ABOUT PROCESS CONTROLLERS

Process control refers to the methods used to maintain the output of process variables– such as temperature, pressure, flow, or level– within a desired range. Precise control of these variables is critical in industrial settings as it improves the quality of products while enabling automation, allowing smaller staffs to monitor and control complex processes from a central location.

Process control is part of a closed loop system in which a process variable is measured, compared to a setpoint, and action is taken to correct any deviation from the setpoint. Closed loop control is feedback-dependent; receiving feedback from sensors monitoring the process variable and providing feedback to the final control element that corrects any deviation from the setpoint. By carefully monitoring and correcting process variables, controllers greatly assist in reducing variability, increasing efficiency, and ensuring safety. Any equipment that requires constant monitoring of a process variable can benefit from a process controller.

Let's use the example of an automated production facility that makes cookies. Process controllers are responsible for delivering a specific ratio of ingredients, mixed them together for an exact amount of time before being portioned into a consistent size and shape. A conveyor transports the raw cookies to the oven where they are baked to a perfect consistency and counted out for packaging.

In the above example, controllers monitor and correct temperature, pressure, batching, humidity and other processes. If any of these were out of specification, the cookies would be ruined. It is the process controllers that reduce variability in the product and guarantee a consistent cookie.

Years ago, workers would've handled all these processes manually-- checking temperatures, mixing ingredients, timing the baking. The process was much slower, less cost effective, and output was lower because of that. Now production is highly automated. It is process controllers that are responsible for the increase in efficiency.

Process variables such as pressure and temperature are potentially dangerous. Many of the "ingredients" used in industry (though not necessarily in cookies) are harmful to people and/or the environment. It is process controllers that maintain conditions to ensure safety.

Control System Components

Process controllers are arranged into control systems (also known as control loops) that consist of the controller, any associated sensors, a power supply, the final control element, as well as any necessary load handling devices.

As we know, controllers seek to maintain the measured process variable at a preset point.

Sensors provide the input signal to the controller. That signal is based upon a measurable physical property like temperature, pressure, pH, flow, level, *etc.* There are a staggering array of sensors, transmitters, and transducers compatible with process controllers. Nearly the only limitation is the type of signal a controller is capable of reading. More sophisticated controllers accept voltage, current, contacts, frequency, thermocouple/RTD, and other signal types.

The final control element refers to the device that acts upon orders from the controller. It can be a heater that is activated when the sensor finds a temperature lower than the set point or a valve that opens when the pressure sensor measures a pressure higher than the set point.

Process controllers, many (though not all) sensors, and final control elements require power to operate. A power supply is an integral element to control loops

Control loops regularly feature additional instruments. Transmitters or signal conditioners are often used to isolate, filter, amplify, or convert a sensor input signal when conditions dictate it. Control loops also frequently include data acquisition devices for archiving information related to the process.

Load handling devices are often needed when the final control element, such as heaters or solenoids, require more power to operate than can be supplied by the controller.

Types of Control Action

Depending upon the unit, process controllers are capable of providing multiple types of control which are suited to different applications and process variables.

On-Off Control

On-off control, also called hysteresis control, is the simplest type of control. As expected, on-off controllers switch abruptly between two states with no middle state. They are for use with equipment that accepts binary input, for example a furnace that is either completely on or completely off.

On-off controllers only switch output when the set point has been crossed. In the case of heating control, the controller switches on when below the set point and off when above the set point. To prevent rapid cycling of the system which can cause damage, hysteresis or on-off differential, is added to the controller operations. The differential prevents cycling by exceeding the setpoint by a small amount before the controller switches on or off.

On-off controllers are often used in applications that don't require precise control, in systems which cannot handle having the energy turned on and off frequently, where the mass of the system is so great that temperatures change extremely slowly, or for temperature alarms.

PID Control

PID control uses three different control terms; proportional (P), integral (I), and derivative (D) to help the controller's algorithms provide a more accurate response to deviations from the set point.

When a controller receives input that a process variable has varied from the set point, instructions are sent to the final control element for correction. For example, a controller receives a signal from a thermocouple that a process temperature is too low prompting the controller to turn on a heater to bring it back up to temperature.

Simple on-off control often leads the final control element to overshoot the set point, especially when the original deviation was small. Repeatedly overshooting the set point causes the output to oscillate around the setpoint in either a constant, growing, or decaying sinusoid. The system is unstable if the amplitude of the oscillations continuously increase with time.

PID controllers use the algorithm derived from their three control terms to maintain system stability by limiting overshoot and resulting oscillation. The proportional variable controls the rate of correction so that it is proportional to the error. The integral and derivative variables are time-based and help the controller automatically compensate to changes in the system. The derivative variable considers the rate at which the error is increasing or decreasing while the integral variable uses knowledge of accumulated errors to the length of time the process is not at the set point. This information is used to correct the proportional value.

PID controllers are generally considered the most efficient type of controller. They are widely used in industrial settings. Though each of the variables must be tuned to a particular system, PID controllers provide very accurate and stable control.

In order to make PID controllers even more responsive to real-world situations, many manufacturers have incorporated **fuzzy logic** (or **fuzzy control**) into the instruments. Fuzzy logic is a mathematical system that attempts to emulate human reasoning. Rather than the binary logic of standard controllers, fuzzy logic introduces continuous variables which provide an effective means of capturing the approximate, inexact nature of the real world.

This ability enables controllers with fuzzy logic to make quick, subtle changes that significantly improves response to fast-changing variables independent of the programming done by the operator. For example, as heaters, valves and other final control elements age, they show signs of wear and no longer respond in the same way as they did when new. Fuzzy logic recognizes this and automatically compensates.

Profile Control

Profile Control refers to controlling a changing process variable against time. Users input the desired time and process profile with the help of extensive instruction set like jump, loop, loop with count apart from ramp and soak control.

Profile control is especially useful for cycling applications which require multiple temperature profiles as well as specific on and off periods.

Limit Control

Limit control involves an independent switch which will shut down the system if a process variable crosses a preset threshold. Limit controllers are for use in processes where for safety or quality issues, a process variable must be kept within specified tolerance levels.

Limit controllers are designed to work in conjunction with another controller. These units also require a manual rest to acknowledge the limit relay has been activated.

FINAL CONTROL ELEMENTS - VARIABLE-SPEED MOTOR CONTROLS

An alternative method of flow control in lieu of control valves is to vary the speed of the machine(s) motivating fluid to flow. In the case of liquid flow control, this would take the form of variable-speed pumps. In the case of gas flow control, it would mean varying the speed of compressors or blowers. Flow control by machine speed control makes a lot of sense for some process applications. It is certainly more energy-efficient to vary the speed of the machine pushing fluid to control flow, as opposed to letting the machine run at full speed all the time and adjusting flow rate by throttling the machine's discharge (outlet) or recycling fluid back to the machine's suction (inlet). The fact that the system has one less component in it (no control valve) also reduces capital investment and potentially increases system reliability:

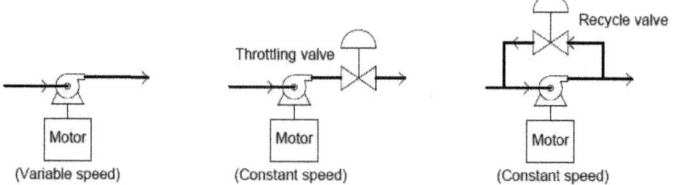

Further advantages of machine speed control include the ability to "soft-start" the machine instead of always accelerating rapidly from a full stop to full speed, reduced wear on machines due to less motion over time, and reduced vibration. In applications such as conveyor belt control, robotic machine motion control, and electric vehicle propulsion, variable-speed technology makes perfect sense because the prime mover device is already (in most cases) an electric motor, with precise speed control of that motor providing many practical benefits. In some applications, regenerative braking may be of benefit, where the motor is used as an electrical generator to slow down the machine on command. Regenerative braking transfers kinetic energy within the machine to the power grid where it may be gainfully used in other processes, saving energy and reducing wear on any mechanical (friction) brakes already installed in the machine.

With all these advantages inherent to variable-speed pumps, fans, and compressors (as opposed to using dissipative control valves), one might wonder, "Why would anyone ever use a control valve to regulate flow? Why not just control all fluid flows using variable-speed pumping machines?" Several good answers exist to this question:

- Variable-speed machines often cannot respond as rapidly as control valves
- Control valves have the ability to positively halt flow; a stopped pump or blower will not necessarily prevent flow from going through
- Some process applications must contain a dissipative element in order for the system to function (*e.g.* let-down valves in closed refrigeration systems)
- Split-ranging may be difficult or impossible to achieve with multiple machine speed control
- Limited options for fail-safe status
- In many cases, there is no machine dedicated to a particular flow path (*e.g.* a pressure release valve)

DC Motor Speed Control

DC electric motors generate torque by a reaction between two magnetic fields: one field established by stationary "field" windings (coils), and the other by windings in the rotating armature. Some DC motors lack field windings, substituting large permanent magnets in their place so that the stationary magnetic field is constant for all operating conditions.

In any case, the operating principle of a DC electric motor is that current passed through the armature creates a magnetic field that tries to align with the stationary magnetic field. This causes the armature to rotate:

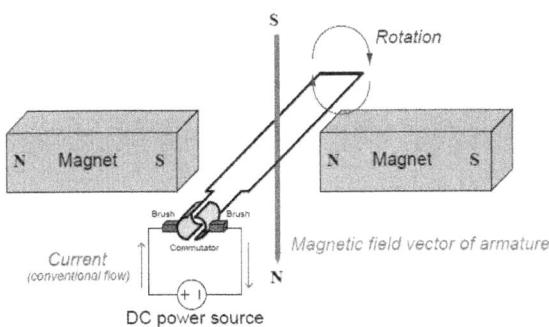

However, a set of segmented copper strips called a commutator breaks electrical contact with the now-aligned coil and energizes another coil (or in the simple example shown above, it re-energizes the same loop of wire in the opposite direction) to create another out-of-alignment magnetic field that continues to rotate the armature. Electrical contact between the rotating commutator segments and the stationary power source is made through carbon brushes. These brushes wear over time (as does the commutator itself), and must be periodically replaced.

Most industrial DC motors are built with multiple armature coils, not just one as shown in the simplified illustration above. A photograph of a large (1250 horsepower) DC motor used to propel a ferry ship is shown here, with the field and armature poles clearly seen (appearing much like spokes in a wheel):

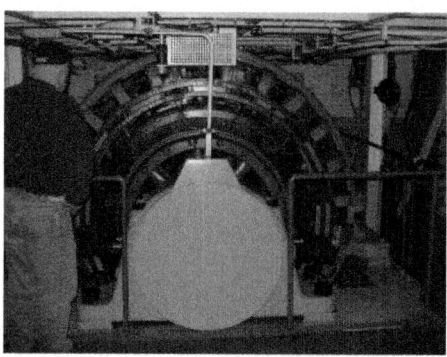

A close-up of one brush assembly on this large motor shows both the carbon brush, the brush's spring-loaded holder, and the myriad of commutator bars the brush makes contact with as the armature rotates:

DC motors exhibit the following relationships between mechanical and electrical quantities:

Torque:

- Torque is directly proportional to armature magnetic field strength, which in turn is directly proportional to current through the armature windings

- Torque is also directly proportional to the stationary pole magnetic field strength, which in turn is directly proportional to current through the field windings (in a motor with non-permanent field magnets)

Speed:

- Speed is limited by the counter-EMF generated by the armature as it spins through the stationary magnetic field. This counter-EMF is directly proportional to armature speed, and also directly proportional to stationary pole magnetic field strength (which is directly proportional to field winding current in a motor that is not permanent-magnet)
- Thus, speed is directly proportional to armature voltage
- Speed is also inversely proportional to stationary magnetic field strength, which is directly proportional to current through the field windings (in a motor with non-permanent field magnets)

A very simple method for controlling the speed and torque characteristics of a wound-field (nonpermanent magnet) DC motor is to control the amount of current through the field winding:

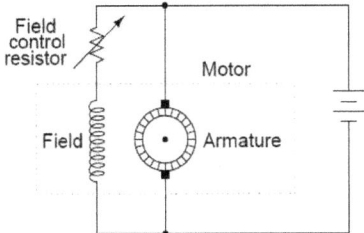

Decreasing the field control resistor's resistance allows more current through the field winding, strengthening its magnetic field. This will have two effects on the motor's operation: first, the motor will generate more torque than it did before (for the same amount of armature current) because there is now a stronger magnetic field for the armature to react against; second, the motor's speed will decrease because more counter-EMF will be generated by the spinning armature for the same rotational speed, and this counter-EMF naturally attempts to equalize with the applied DC source voltage. Conversely, we may increase a DC motor's speed (and reduce its torque output) by increasing the field control resistor's resistance, weakening the stationary magnetic field through which the armature spins.

Regulating field current may alter the balance between speed and torque, but it does little to control total motor power. In order to control the power output of a DC motor, we must also regulate armature voltage and current. Variable resistors may also be used for this task, but this is generally frowned upon in modern times because of the wasted power.

A better solution is to have an electronic power control circuit very rapidly switch transistors on and off, switching power to the motor armature. This is called pulse-width modulation, or PWM.

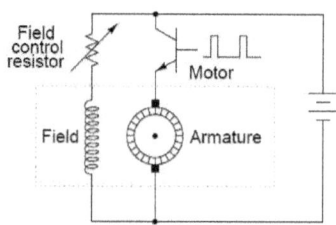

The duty cycle (on time versus on+off time) of the pulse waveform will determine the fraction of total power delivered to the motor:

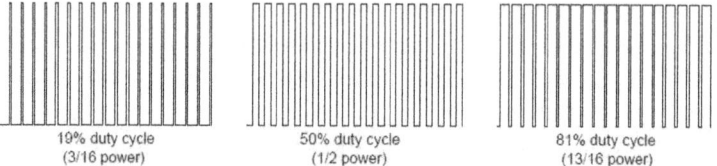

| 19% duty cycle | 50% duty cycle | 81% duty cycle |
| (3/16 power) | (1/2 power) | (13/16 power) |

Such an electronic power-control circuit is generally referred to as a drive. Thus, a variable-speed drive or VSD is a high-power circuit used to control the speed of a DC motor. Motor drives may be manually set to run a motor at a set speed, or accept an electronic control signal to vary the motor speed in the same manner an electronic signal commands a control valve to move. When equipped with remote control signaling, a motor drive functions just like any other final control element: following the command of a process controller in order to stabilize some process variable at setpoint.

An older technology for pulsing power to a DC motor is to use a controlled rectifier circuit, using SCRs instead of regular rectifying diodes to convert AC to DC. Since the main power source of most industrial DC motors is AC anyway, and that AC must be converted into DC at some point in the system, it makes sense to integrate control right at the point of rectification:

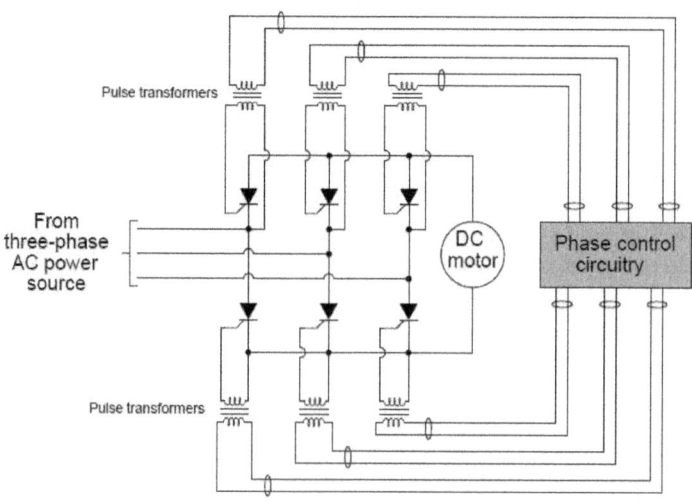

Controlled rectifier circuits work on the principle of varying the "trigger" pulse times relative to the AC waveform pulses. The earlier the AC cycle each SCR is triggered on, the longer it will be on to pass current to the motor. The "phase control" circuitry handles all this pulse timing and generation.

A DC motor drive that simply varied power to the motor according to a control signal would be crude and difficult to apply to the control of most processes. What is ideally desired from a variable-speed drive is precise command over the motor's speed. For this reason, most VSDs are designed to receive feedback from a tachometer mechanically connected to the motor shaft, so the VSD "knows" how fast the motor is turning. The tachometer is typically a small DC generator, producing a DC voltage directly proportional to its shaft speed (0 to 10 volts is a common scale). With this information, the VSD may throttle electrical power to the motor as necessary to achieve whatever speed is being commanded by the control signal. Having a speed-control feedback loop built into the drive makes the VSD a "slave controller" in a cascade control system, the drive receiving a speed setpoint signal from whatever process controller is sending an output signal to it:

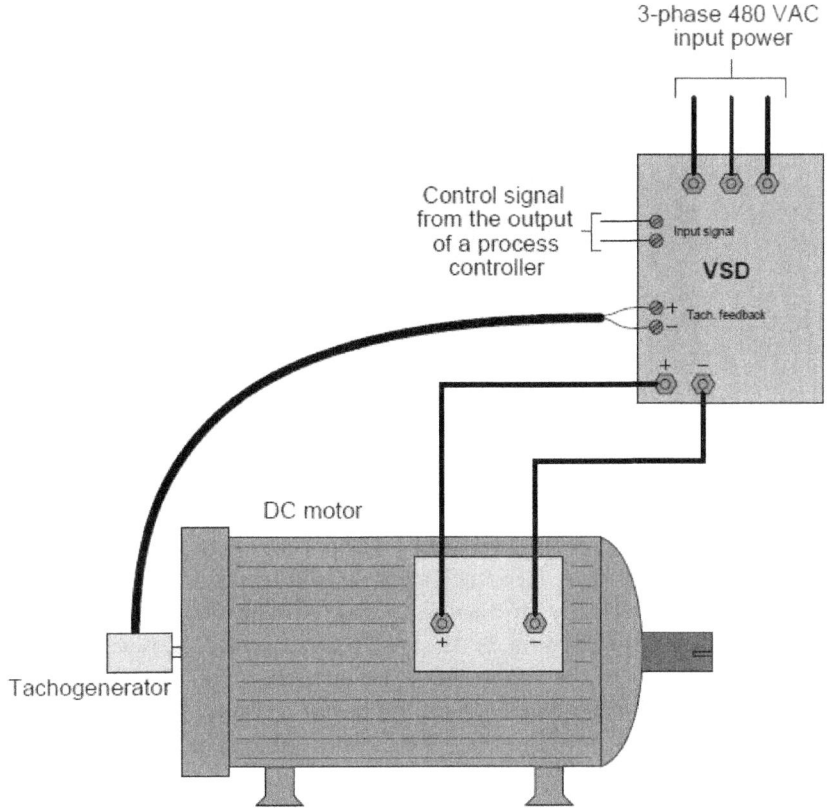

A photograph of the tachogenerators (dual, for redundancy) mechanically coupled to that large 1250 horsepower ferry ship propulsion motor appears here:

The SCRs switching power to this motor may be seen here, connected *via* twisted-pair wires to control boards issuing "firing" pulses to each SCR at the appropriate times:

The integrity of the tachogenerator feedback signal to the VSD is extremely important for safety reasons. If the tachogenerator becomes disconnected – whether mechanically or electrically (it doesn't matter) – from the drive, the drive will "think" the motor is not turning. In its capacity as a speed controller, the drive will then send full power to the DC motor in an attempt to get it up to speed. Thus, loss of tachogenerator feedback causes the motor to immediately "run away" to full speed. This is undesirable at best, and likely dangerous in the case of motors as large as the one powering this ship.

As with all forms of electric power control based on pulse durations and duty cycles, there is a lot of electrical "noise" cast by VSD circuits. Square-edged pulse waveforms created by the rapid on-and-off switching of the semiconductor power devices are equivalent to infinite series of high-frequency sine waves, some of which may be of high enough frequency to self-propagate through space as electromagnetic waves. This radio-frequency interference or RFI may be quite severe given the high power levels of industrial motor drive circuits. For this reason, it is imperative that neither the motor power conductors nor the conductors feeding AC power to the drive circuit be routed anywhere near small-signal or control wiring, because the induced noise will wreak havoc with whatever systems utilize those low-level signals.

RFI noise on the AC power conductors may be mitigated by routing the AC power through filter circuits placed near the drive. The filter circuits block high-frequency noise from propagating back to the rest of the AC power distribution wiring where it may influence other electronic equipment. However, there is little that may be done about the RFI noise between the drive and the motor other than to shield the conductors in well-grounded metallic conduit.

AC Motor Speed Control

AC induction motors are based on the principle of a rotating magnetic field produced by a set of stationary windings (called stator windings) energized by AC power of different phases. The effect is not unlike a series of blinking light bulbs which appear to "move" in one direction due to intentional sequencing of the blinking. If sets of wire coils (windings) are energized in a like manner – each coil reaching its peak field strength at a different time from its adjacent neighbor – the effect will be a magnetic field that "appears" to move in one direction. If these windings are oriented around the circumference of a circle, the moving magnetic field rotates about the center of the circle, as illustrated by this sequence of images (read left-to-right, top-to-bottom, as if you were reading words in a sentence):

Any magnetized object placed in the center of this circle will attempt to spin at the same rotational speed as the rotating magnetic field. Synchronous AC motors use this principle, where a magnetized rotor precisely follows the magnetic field's speed.

Any electrically conductive object placed in the center of the circle will experience induction as the magnetic field direction changes around the conductor. This will induce electric currents within the conductive object, which in turn will react against the rotating magnetic field in such a way that the object will be "dragged along" by the field, always lagging a bit in speed. Induction AC motors use this principle, where a non-magnetized (but electrically conductive) rotor rotates at a speed slightly less than the synchronous speed of the rotating magnetic field.

The rotational speed of this magnetic field is directly proportional to the frequency of the AC power, and inversely proportional to the number of poles in the stator:

$$S = \frac{120f}{n}$$

Where,

S = Synchronous speed of rotating magnetic field, in revolutions per minute (RPM)

f = Frequency, in cycles per second (Hz)

n = Total number of stator poles per phase (the simplest possible AC motor design will have 2 poles)

While the number of poles in the motor's stator is a quantity fixed at the time of the motor's manufacture, the frequency of power we apply may be adjusted with the proper electronic circuitry. A high-power circuit designed to produce varying frequencies for an AC motor to run on is called a variable-frequency drive, or VFD.

A simplified schematic diagram for a VFD is shown here, with a rectifier section on the left (to convert AC input power into DC), a filter capacitor to "smooth" the rectified DC power, and a transistor "bridge" to switch DC into AC at whatever frequency is desired to power the motor. The transistor control circuitry has been omitted from this diagram for the sake of simplicity:

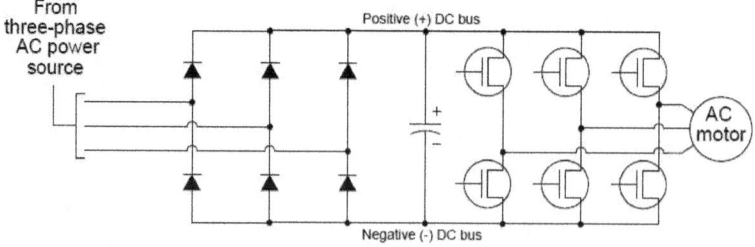

As with DC motor drives (VSDs), the power transistors in an AC drive (VFD) switch on and off very rapidly with a varying duty cycle. Unlike DC drives, however, the duty cycle of an AC drive's power transistors must vary rapidly in order to synthesize an AC waveform from the DC "bus" voltage following the rectifier. A DC drive circuit's PWM duty cycle controls motor power, and so it will remain at a constant value when the desired motor power is constant. Not so for an AC motor drive circuit: its duty cycle must vary from zero to maximum and back to zero repeatedly in order to generate an AC waveform for the motor to run on.

The equivalence between a rapidly-varied pulse-width modulation (PWM) waveform and a sine wave is shown in the following illustration:

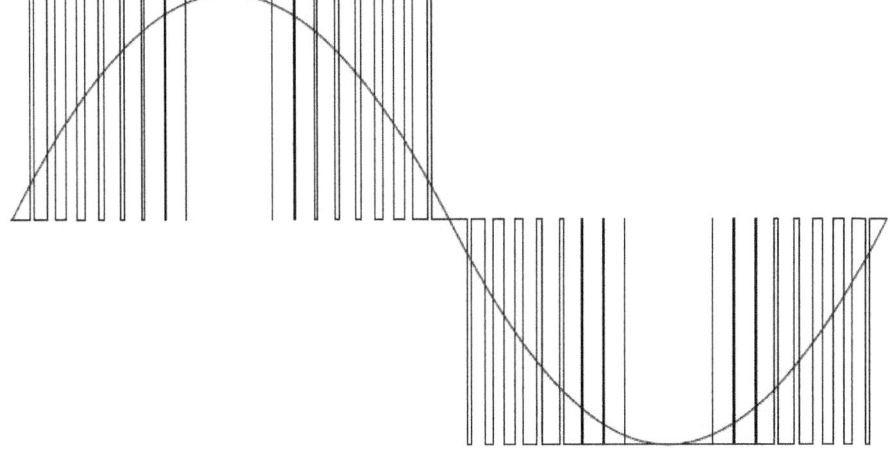

This concept of rapid PWM transistor switching allows the drive to "carve" any arbitrary waveform out of the filtered DC voltage it receives from the rectifier. Virtually any frequency may be synthesized (up to a maximum limited by the frequency of the PWM pulsing), and any voltage (up to a maximum peak established by the DC bus voltage), giving the VFD the ability to power an induction motor over a wide range of speeds.

While frequency control is the key to synchronous and induction AC motor speed control, it is generally not enough on its own. While the speed of an AC motor is a direct function of frequency (controlling how fast the rotating magnetic field rotates around the circumference of the stator), torque is approximately proportional to stator winding current. Since the stator windings are inductors by nature, their reactance varies with frequency as described by the formula $X_L = 2\pi f L$. Thus, as frequency is increased, winding reactance increases right along with it. This increase in reactance results in a decreased stator current (assuming the AC voltage is held constant as frequency is increased). This can cause undue torque loss at high speeds, and excessive torque (as well as excessive stator heat!) at low speeds. For this reason, the AC voltage applied to a motor by a VFD is usually made to vary in direct proportion to the applied frequency, so that the stator current will remain within good operating limits throughout the speed range of the VFD. This correspondence is called the voltage-to-frequency ratio, or V/F ratio.

Variable-frequency motor drives are manufactured for industrial motor control in a wide range of sizes and horsepower capabilities. Some VFDs are small enough to hold in your hand, while others are large enough to require a freight train for transport. The following photograph shows a pair of moderately-sized Allen-Bradley VFDs (about 100 horsepower each, standing about 4 feet high), used to control pumps at a wastewater treatment plant:

Variable-frequency AC motor drives do not require motor speed feedback the way variable-speed DC motor drives do. The reason for this is quite simple: the controlled variable in an AC drive is the frequency of power sent to the motor, and rotating-magnetic-field AC motors are frequency-controlled machines by their very nature. For example, a 4-pole AC induction motor powered by 60 Hz has a base speed of 1728 RPM (assuming 4% slip). If a VFD sends 30 Hz AC power to this motor, its speed will be approximately half its base-speed value, or 864 RPM. There is really no need for speed-sensing feedback in an AC drive, because the motor's real speed will always be limited by the drive's output frequency. To control frequency is to control motor speed for AC synchronous and induction motors, so no tachogenerator feedback is necessary for an AC drive to "know" approximately how fast the motor is turning. The non-necessity of speed feedback for AC drives eliminates a potential safety hazard common to DC drives: the possibility of a "runaway" event where the drive loses its speed feedback signal and sends full power to the motor.

As with DC motor drives, there is a lot of electrical "noise" cast by VFD circuits. Square-edged pulse waveforms created by the rapid on-and-off switching of the power transistors are equivalent to infinite series of high-frequency sine waves, some of which may be of high enough frequency to self-propagate through space as electromagnetic waves. This radio-frequency interference or RFI may be quite severe given the high power levels of industrial motor drive circuits. For this reason, it is imperative that neither the motor power conductors nor the conductors feeding AC power to the drive circuit be routed anywhere near small-signal or control wiring, because the induced noise will wreak havoc with whatever systems utilize those low-level signals.

RFI noise on the AC power conductors may be mitigated by routing the AC power through filter circuits placed near the drive. The filter circuits block high-frequency noise from propagating back to the rest of the AC power distribution wiring where it may influence other electronic equipment. However, there is little that may be done about the RFI noise between the drive and the motor other than to shield the conductors in well-grounded metallic conduit.

Motor Drive Features

Modern DC and AC motor drives provide features useful when using electric motors as final control elements. Some common features seen in both VSDs and VFDs are listed here:

- Speed limiting
- Torque limiting
- Torque profile curves (used to regulate the amount of torque available at different motor speeds)
- Acceleration (speed rate-of-change) limiting
- Deceleration (speed rate-of-change) limiting
- Dynamic braking (turning the motor into an electromagnetic brake)
- Plugging (applying reverse-direction power to a motor to quickly stop it)
- Regenerative braking (turning the motor into a generator to recover kinetic energy)
- Overcurrent monitoring and automatic shut-down
- Overvoltage monitoring and automatic shut-down
- PWM frequency adjustment (may be helpful in reducing electromagnetic interference with some equipment)

Not only are some of these limiting parameters useful in extending the life of the motor, but they may also help extend the operating life of the mechanical equipment powered by the motor. It is certainly advantageous, for example, to have torque limiting on a conveyor belt motor, so that the motor does not apply full rated torque (*i.e.* stretching force) to the belt during start-up.

If a motor drive is equipped with digital network communication capability (*e.g.* Modbus), it is usually possible for a host system such as a PLC or DCS to update these control parameters as the motor is running.

Metering Pumps

A very common method for directly controlling low flow rates of fluids is to use a device known as a metering pump. A "metering pump" is a pump mechanism, motor, and drive electronics contained in a monolithic package. Simply supply 120 VAC power and a control signal to a metering pump, and it is ready to use.

Metering pumps are commonly used in water treatment processes to inject small quantities of treatment chemicals (*e.g.* coagulants, disinfectants, acid or caustic liquids for pH neutralization, corrosion-control chemicals) into the water flowstream, as is the Milton-Roy unit shown in this photograph:

Adjustment knobs on the front of the pump establish the maximum flow rate at a control signal value of 100%:

While some metering pumps use rotary motor and pump mechanisms, many use a "plunger" style mechanism operated by a solenoid at variable intervals. Thus, the latter type of metering pump does not provide continuous flow control, but rather a flow consisting of discrete pulse events distributed over a period of time. The "plunger" metering pumps are quite simple and reliable, and are entirely appropriate if non-continuous flow is permissible for the process.

UNDERSTANDING TIME PROPORTIONAL CONTROL

An underused control strategy that offers significant benefits is Time-Proportional Control (TPC). Unlike traditional proportional or even PID control that require a varying output to a modulating control device, time-proportional control can achieve a proportional control response to process variation using an on/off device by varying on and off times in a defined control period. The on/off device is generally a simpler, less expensive control device.

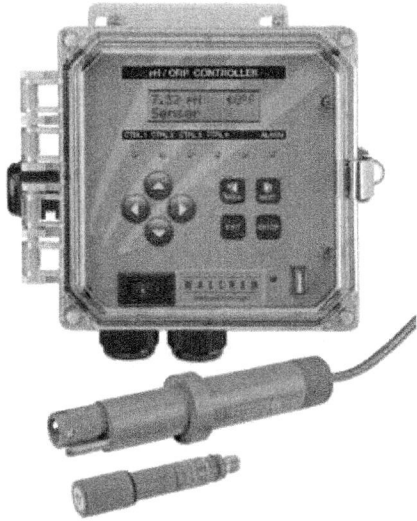

Walchem's WPH Controller offers software algorithms for time-proportional control.

First, a quick review of the first two common methods. Proportional analog is the most common control signal. This signal is often sent to a modulating control valve (positioner), variable speed pump (centrifugal or diaphragm metering), or to a variety of mechanical control options such as speed of a belt drive, *etc.* The premise is straightforward – the signal value from 4-20 mA creates a direct, proportional output response (0-100%) of the control device.

In basic proportional only control, the signal value is calculated from a linear relationship based on the variance from set-point with a maximum output set at a fixed maximum variance. A common example is in pH control where the output to an acid metering pump will vary from 4-20 mA as the pH increases from 8-10 pH. At 8 pH or less, 4 mA is sent to achieve 0% output of the pump. As the pH increases to 10 pH, the output will increase linearly, until at 10 or greater pH, 20 mA will be sent to the pump to achieve 100% output.

In more sophisticated control scenarios, the algorithm to determine the output value may incorporate Integral and possibly Derivative calculations (PI, or PID Control) to enable a more predictive output based on the speed of the process response to the output. Even with PI or PID control, the actual output signal remains a 4-20mA resulting in a proportional 0-100% response of the control device.

Proportional Pulse control is used specifically for proportional control of solenoid driven metering pumps. A controller varies a pulse output based on deviation from setpoint, similar to Proportional analog control. Using our same acid feed control example as above, the controller will send 0 pulses per minute to the metering pump at pH 8 or less, increasing the pulse/minute output proportional until at pH 10 or greater, the maximum pulse output will be generated. The metering pump will stroke one time for each pulse it receives, enabling the increase in pH to cause an increase in the volume output of the metering pump.

Time-Proportional Control

Time-Proportional control is a less widely used method for achieving proportional control, and has the advantage that it uses a lower cost on/off control device such as a solenoid valve, or fixed output pump. By proportioning the on-time *vs.* off-time of the control device within a fixed time period (sample period), a proportional response is achieved. The off-time portions of control provide an additional benefit by enabling better mixing of the process to occur, or time for reactions to take place.

The parameters used to program a time-proportional output include the sample period, the set-point, the proportional band, and the control direction. The set-point is the desired pH of the system; while the control direction determines whether the output will increase above the set point (often called a High Set-Point) or below the set-point (Low Set-Point).

The sample period should be set to approximately 1½ times the amount of time that it takes for the system to react to the chemical addition. This can be determined by making a manual addition of chemical and timing how long it takes for the process to react. Setting the sample period too low will result in a second addition being made before the first is detected and will cause set point overshoot. Setting the sample period too high will delay the next addition and can prevent the set point being reached.

Finally, the Proportional Band is the deviation from set point that will result in a 100% on time of the output. Returning to our acid feed pump example for pH control, we could set a sample period of 10 minutes, a High Set-Point of 8.0, and a Proportional Band of 2 pH (from 8 to 10). At 8 pH or below, the output would be off 100% of the time. At pH 10.0 or above, the output would be on for 100% of the next 10 minute sample period. At 9 pH, the output would be on for 5 minutes, then off for 5 minutes. The percentage of on-time of the 10 minute sample period increases proportionally as the process moves away from the setpoint. The actual on-time can be calculated as follows:

$$\frac{\text{Actual pH} - \text{Set Point}}{\text{Proportional Band}} * \text{Sample Period} = \text{On-time of Device}$$

The off-time will be the remainder of the Sample Period time.

The set point is 8.0 pH and the Proportional Band is 2.0 pH. Note that when the pH goes above the set point, the control relay is ON for a short period of time. As the pH increases, the control relay is ON for a longer period of time. When the addition starts to affect the process pH and the pH is reduced, the control relay is ON for a shorter period of time. When the pH drops below the set point of 8.0, the control relay is OFF all the time.

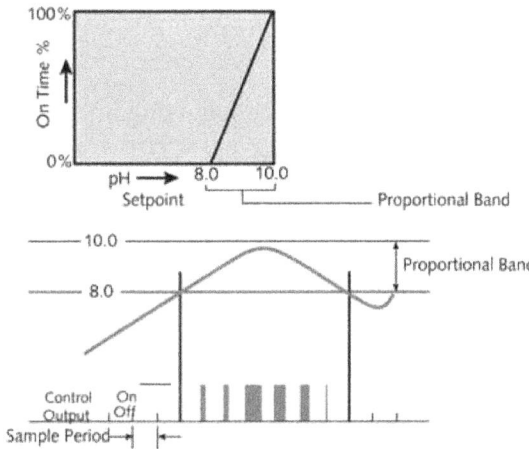

Fig. : Graphic Representation of Time-Proportional Control for pH.

In addition to lower cost control devices, there can also be other process benefits. Some plants will use waste process streams as their neutralizing chemicals (*i.e.* waste acid baths as acid reagent, or waste cleaners (generally alkaline) as caustic reagents). Many of these streams are too dirty to be pumped using standard solenoid metering pumps. A solenoid or actuated ball or diaphragm valve on a gravity feed system can provide a lower maintenance option.

Time-proportional control should not be used in once through (on-the-fly) control processes, such as chemical injection into a pipe. These systems require feedback and prediction algorithms that are found in PID. Time-proportional control is well suited for recirculating processes, or those with a built in retention time (*i.e.* neutralization tank). A system with a retention time of 10 minutes or longer is suitable. To determine the retention time of a system, simply divide the tank volume by the flow through the tank, *i.e.* a 600 gallon tank with flow in/out at 60 gpm equals a 10 minute retention time.

Time-proportional control has been available for a long time, but is often overlooked by process control experts who tend to use more sophisticated PID algorithms, often with self-tuning or other enhancements, even when they're not necessary. While there is a place for these, the simpler time-proportional control algorithm should not be forgotten. It can provide simple, enhanced control using lower maintenance components to achieve improved control in many applications.

PID CONTROLLER DESIGN

In this tutorial we will introduce a simple yet versatile feedback compensator structure, the Proportional-Integral-Derivative (PID) controller. The effect of each of the pid parameters on the closed-loop dynamics and demonstrate how to use a PID controller to improve the system performance.

Key MATLAB commands used in this tutorial are: `tf`, `step`, `pid`, `feedback`, `pidtool`, `pidtune`

PID Overview

In this tutorial, we will consider the following unity feedback system:

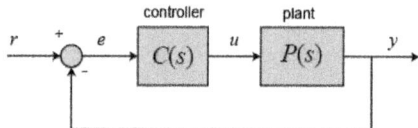

The output of a PID controller, equal to the control input to the plant, in the time-domain is as follows:

$$(1)\quad u(t) = K_p e(t) + K_i \int e(t) dt + K_p \frac{de}{dt}$$

First, let's take a look at how the PID controller works in a closed-loop system using the schematic shown above. The variable (e) represents the tracking error, the difference between the desired input value (r) and the actual output (y). This error signal (e) will be sent to the PID controller, and the controller computes both the derivative and the integral of this error signal. The control signal (u) to the plant is equal to the proportional gain (K_p) times the magnitude of the error plus the integral gain (K_i) times the integral of the error plus the derivative gain (K_d) times the derivative of the error.

This control signal (u) is sent to the plant, and the new output (y) is obtained. The new output (y) is then fed back and compared to the reference to find the new error signal (e). The controller takes this new error signal and computes its derivative and its integral again, ad infinitum.

The transfer function of a PID controller is found by taking the Laplace transform of Eq.(1).

$$(2)\quad K_p + \frac{K_i}{s} + K_d s = \frac{K_d s^2 + K_p s + K_i}{s}$$

K_p = Proportional gain K_i = Integral gain K_d = Derivative gain

We can define a PID controller in MATLAB using the transfer function directly, for example:

```
Kp = 1;
Ki = 1;
```

```
Kd = 1;
s = tf('s');
C = Kp + Ki/s + Kd*s
C =
s^2 + s + 1
-----------
      s
```

Continuous-time transfer function.

Alternatively, we may use MATLAB's pid controller object to generate an equivalent continuous-time controller as follows:

```
C = pid(Kp,Ki,Kd)
C =
              1
  Kp + Ki * --- + Kd * s
              s
  with Kp = 1, Ki = 1, Kd = 1
```

Continuous-time PID controller in parallel form.

Let's convert the pid object to a transfer function to see that it yields the same result as above:

```
tf(C)
ans =

  s^2 + s + 1
  -----------
        s
```

Continuous-time transfer function.

The Characteristics of P, I, and D Controllers

A proportional controller (K_p) will have the effect of reducing the rise time and will reduce but never eliminate the steady-state error. An integral control (K_i) will have the effect of eliminating the steady-state error for a constant or step input, but it may make the transient response slower. A derivative control (K_d) will have the effect of increasing the stability of the system, reducing the overshoot, and improving the transient response.

The effects of each of controller parameters, K_p, K_d, and K_i on a closed-loop system are summarized in the table below.

CL Response	Rise Time	Overshoot	Settling Time	S-S Error
K_p	Decrease	Increase	Small Change	Decrease
K_i	Decrease	Increase	Increase	Eliminate
K_d	Small Change	Decrease	Decrease	No Change

Note that these correlations may not be exactly accurate, because K_p, K_i, and K_d are dependent on each other. In fact, changing one of these variables can change the effect of the other two. For this reason, the table should only be used as a reference when you are determining the values for K_i, K_p and K_d.

Example Problem

Suppose we have a simple mass, spring, and damper problem.

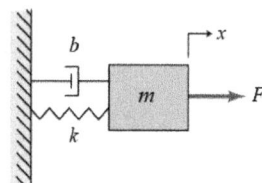

The modeling equation of this system is

(3) $M\ddot{x} + b\dot{x} + kx = F$

Taking the Laplace transform of the modeling equation, we get

(4) $Ms^2 X(s) + bsX(s) + kX(s) = F(s)$

The transfer function between the displacement $X(s)$ and the input $F(s)$ then becomes

(5) $\dfrac{X(s)}{F(s)} = \dfrac{1}{Ms^2 + bs + k}$

Let

```
M = 1 kg
b = 10 N s/m
k = 20 N/m
F = 1 N
```

Plug these values into the above transfer function

(6) $\dfrac{X(s)}{F(s)} = \dfrac{1}{s^2 + 10s + 20}$

The goal of this problem is to show you how each of K_p, K_i and K_d contributes to obtain

```
Fast rise time
Minimum overshoot
No steady-state error
```

Open-Loop Step Response

Let's first view the open-loop step response. Create a new m-file and run the following code:

```
s = tf('s');
P = 1/(s^2 + 10*s + 20);
step(P)
```

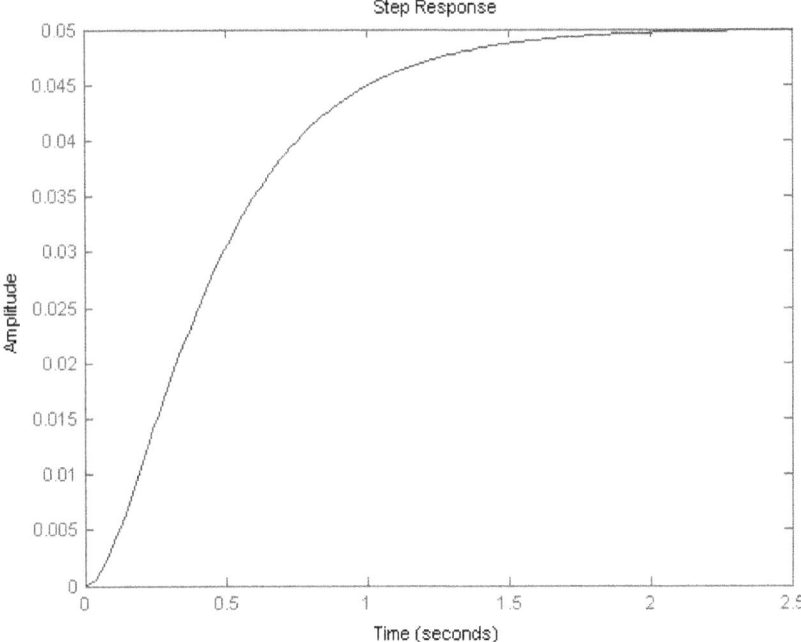

The DC gain of the plant transfer function is 1/20, so 0.05 is the final value of the output to an unit step input. This corresponds to the steady-state error of 0.95, quite large indeed. Furthermore, the rise time is about one second, and the settling time is about 1.5 seconds. Let's design a controller that will reduce the rise time, reduce the settling time, and eliminate the steady-state error.

Proportional Control

From the table shown above, we see that the proportional controller (Kp) reduces the rise time, increases the overshoot, and reduces the steady-state error.

The closed-loop transfer function of the above system with a proportional controller is:

$$(7) \quad \frac{X(s)}{F(s)} = \frac{K_p}{s^2 + 10s + (20 + K_p)}$$

Let the proportional gain (K_p) equal 300 and change the m-file to the following:

```
Kp = 300;
C = pid(Kp)
T = feedback(C*P,1)
t = 0:0.01:2;
step(T,t)
C =

    Kp = 300

  P-only controller.


T =

            300
    ----------------
    s^2 + 10 s + 320
Continuous-time transfer function.
```

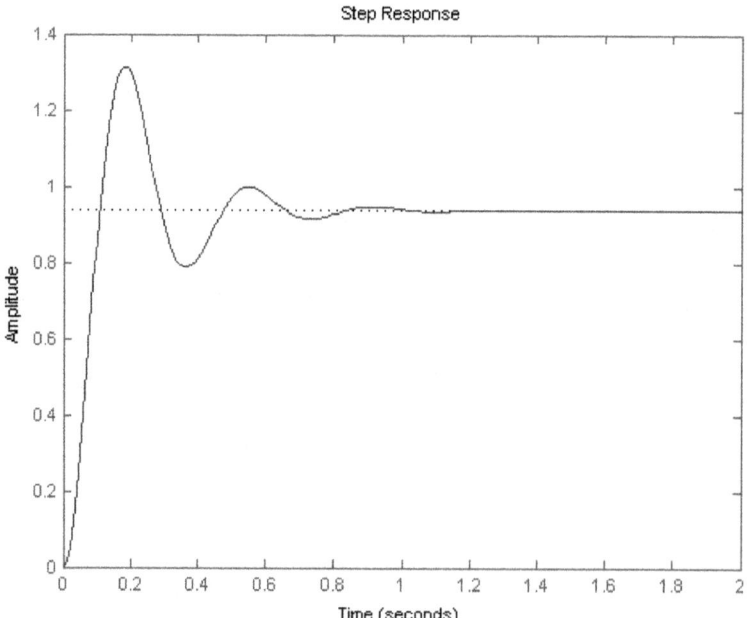

The above plot shows that the proportional controller reduced both the rise time and the steady-state error, increased the overshoot, and decreased the settling time by small amount.

Proportional-Derivative Control

Now, let's take a look at a PD control. From the table shown above, we see that the derivative controller (K_d) reduces both the overshoot and the settling time. The closed-loop transfer function of the given system with a PD controller is:

$$(8) \quad \frac{X(s)}{F(s)} = \frac{K_d s + K_p}{s^2 + (10 + K_d)s + (20 + K_p)}$$

Let K_p equal 300 as before and let K_d equal 10. Enter the following commands into an m-file and run it in the MATLAB command window.

```
Kp = 300;
Kd = 10;
C = pid(Kp,0,Kd)
T = feedback(C*P,1)
t = 0:0.01:2;
step(T,t)
C =

  Kp + Kd * s
  with Kp = 300, Kd = 10
Continuous-time PD controller in parallel form.
T =

      10 s + 300
   ------------------
   s^2 + 20 s + 320
Continuous-time transfer function.
```

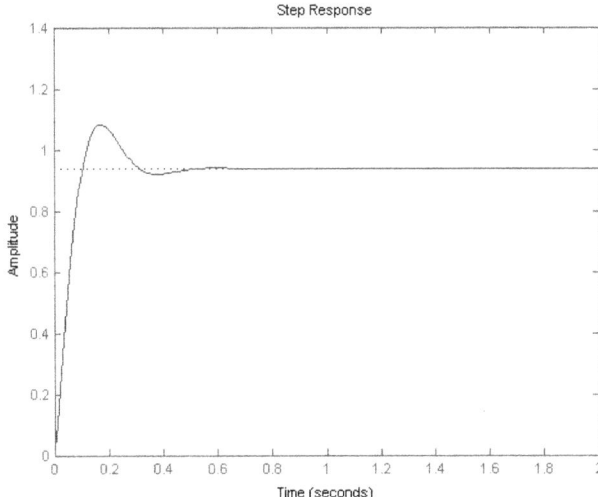

This plot shows that the derivative controller reduced both the overshoot and the settling time, and had a small effect on the rise time and the steady-state error.

Proportional-Integral Control

Before going into a PID control, let's take a look at a PI control. From the table, we see that an integral controller (Ki) decreases the rise time, increases both the overshoot and the settling time, and eliminates the steady-state error. For the given system, the closed-loop transfer function with a PI control is:

$$(9) \quad \frac{X(s)}{F(s)} = \frac{K_p s + K_i}{s^2 + 10s^2 + (20 + K_p s + K_i)}$$

Let's reduce the K_p to 30, and let K_i equal 70. Create an new m-file and enter the following commands.

```
Kp = 30;

Ki = 70;

C = pid(Kp,Ki)

T = feedback(C*P,1)

t = 0:0.01:2;

step(T,t)

C =

                1

   Kp + Ki *  ---

                s

   with Kp = 30,  Ki = 70

Continuous-time PI controller in parallel form.

T =


             30 s + 70

   -------------------------

   s^3 + 10 s^2 + 50 s + 70


Continuous-time transfer function.
```

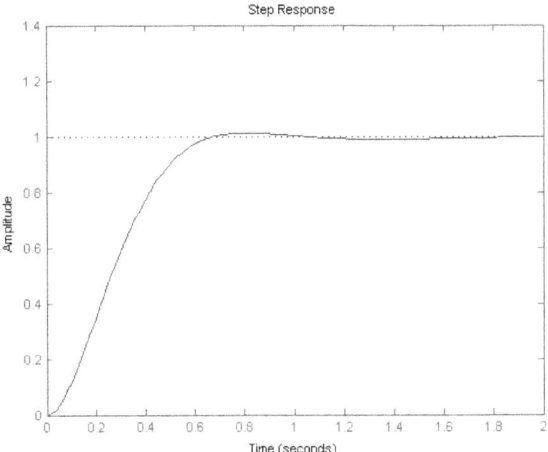

Run this m-file in the MATLAB command window, and you should get the following plot. We have reduced the proportional gain (Kp) because the integral controller also reduces the rise time and increases the overshoot as the proportional controller does (double effect). The above response shows that the integral controller eliminated the steady-state error.

Proportional-Integral-Derivative Control

Now, let's take a look at a PID controller. The closed-loop transfer function of the given system with a PID controller is:

$$(10)\quad \frac{X(s)}{F(s)} = \frac{K_p s^2 + K_p s + K_i}{s^3 + (10 + K_d)s^2 + (20 + K_p)s + K_i}$$

After several trial and error runs, the gains $K_p = 350$, $K_i = 300$, and $K_d = 50$ provided the desired response. To confirm, enter the following commands to an m-file and run it in the command window. You should get the following step response.

```
Kp = 350;
Ki = 300;
Kd = 50;
C = pid(Kp,Ki,Kd)
T = feedback(C*P,1);
t = 0:0.01:2;
step(T,t)
C =

              1
Kp + Ki * --- + Kd * s

              s
with Kp = 350, Ki = 300, Kd = 50
Continuous-time PID controller in parallel form.
```

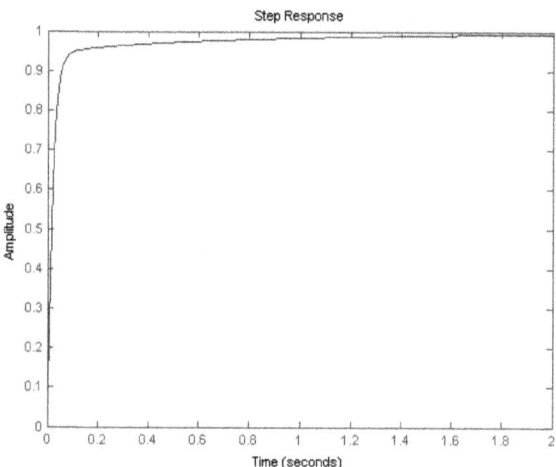

Now, we have obtained a closed-loop system with no overshoot, fast rise time, and no steady-state error.

General Tips for Designing a PID Controller

When you are designing a PID controller for a given system, follow the steps shown below to obtain a desired response.

1. Obtain an open-loop response and determine what needs to be improved
2. Add a proportional control to improve the rise time
3. Add a derivative control to improve the overshoot
4. Add an integral control to eliminate the steady-state error
5. Adjust each of Kp, Ki, and Kd until you obtain a desired overall response. You can always refer to the table shown in this "PID Tutorial" page to find out which controller controls what characteristics.

Lastly, please keep in mind that you do not need to implement all three controllers (proportional, derivative, and integral) into a single system, if not necessary. For example, if a PI controller gives a good enough response (like the above example), then you don't need to implement a derivative controller on the system. Keep the controller as simple as possible.

Automatic PID Tuning

MATLAB provides tools for automatically choosing optimal PID gains which makes the trial and error process described above unnecessary. You can access the tuning algorithm directly using pidtune or through a nice graphical user interface (GUI) using pidtool.

The MATLAB automated tuning algorithm chooses PID gains to balance performance (response time, bandwidth) and robustness (stability margins). By default the algorthm designs for a 60 degree phase margin.

Let's explore these automated tools by first generating a proportional controller for the mass-spring-damper system by entering the following commands:

```
pidtool(P,'p')
```

The pidtool GUI window, like that shown below, should appear.

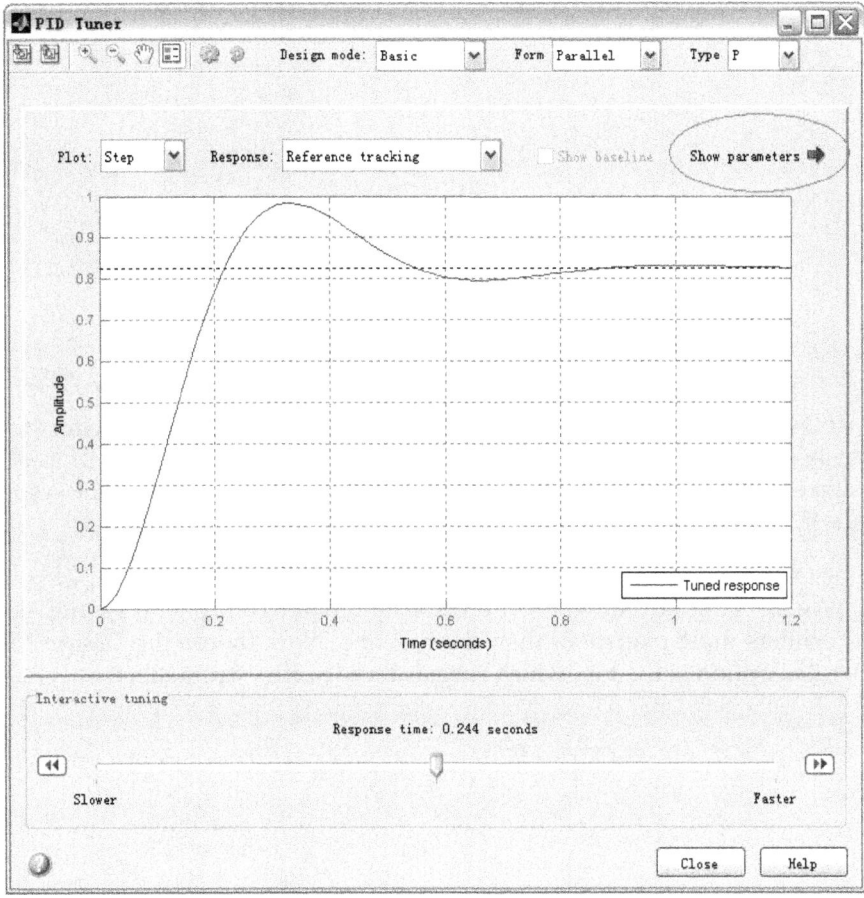

Notice that the step response shown is slower than the proportional controller we designed by hand. Now click on the Show Parameters button on the top right. As expected the proportional gain constant, K_p, is lower than the one we used, $K_p = 94.85 < 300$.

We can now interactively tune the controller parameters and immediately see the resulting response int he GUI window. Try dragging the resposne time slider to the right to 0.14s. The response does indeeed speed up, and we can see K_p is now closer to the manual value. We can also see all the other performance and robustness parameters for the system. Note that the phase margin is 60 degrees, the default for pidtool and generally a good balance of robustness and performance.

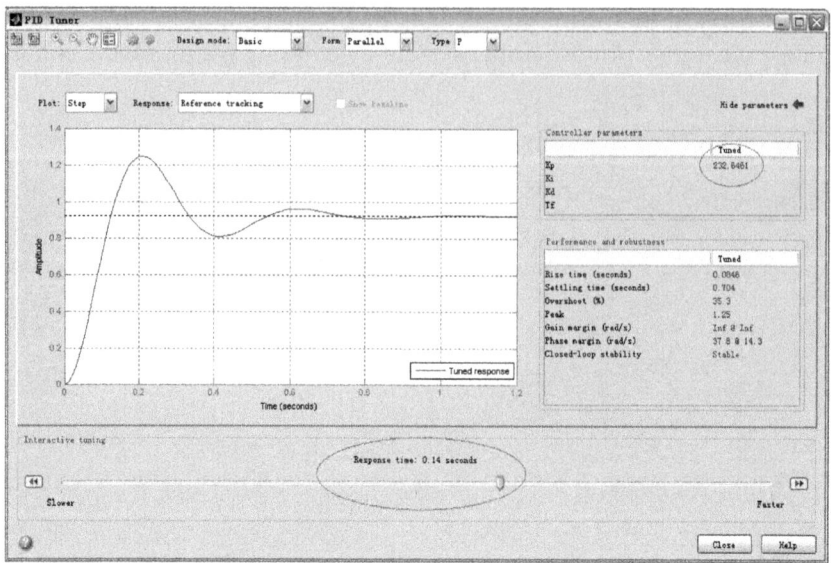

Now let's try designing a PID controller for our system. By specifying the previously designed or (baseline) controller, C, as the second parameter, pidtool will design another PID controller (instead of P or PI) and will compare the response of the system with the automated controller with that of the baseline.

```
pidtool(P,C)
```

We see in the output window that the automated controller responds slower and exhibits more overshoot than the baseline. Now choose the Design Mode: Extended option at the top, which reveals more tuning parameters.

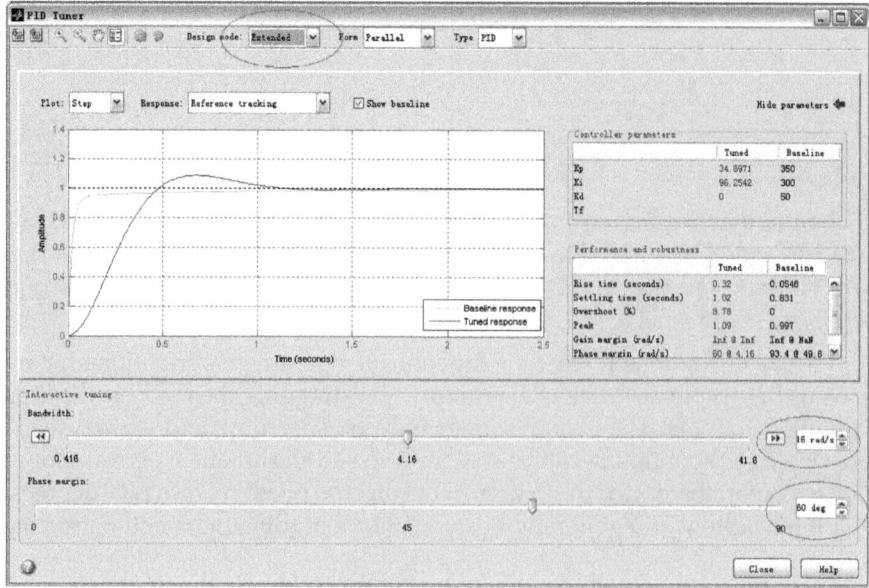

Now type in Bandwidth: 32 rad/s and Phase Margin: 90 deg to generate a controller similar in performance to the baseline. Keep in mind that a higher bandwidth (0 dB crossover of the open-loop) results in a faster rise time, and a higher phase margin reduces the overshoot and improves the system stability.

ANATOMY OF A FEEDBACK CONTROL SYSTEM

Here is the classic block diagram of a process under PID Control.

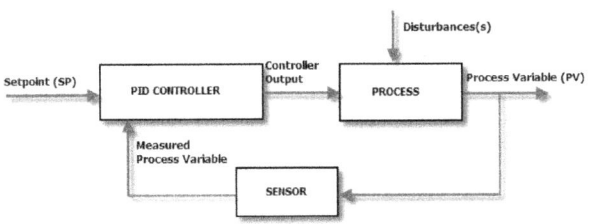

What's going on this diagram?

The **Setpoint** (SP) is the value that we **want** the process to be.

For example, the temperature control system in our house may have a SP of 22°C. This means that

"we want the heating and cooling process in our house to achieve a steady temperature of as close to 22°C as possible"

The PID controller looks at the setpoint and compares it with the actual value of the **Process Variable (PV)**. Back in our house, the box of electronics that is the PID controller in our Heating and Cooling system looks at the value of the temperature sensor in the room and sees how close it is to 22°C.

If the SP and the PV are the same – then the controller is a very happy little box. It doesn't have to do anything, it will set its output to zero.

However, if there is a disparity between the SP and the PV we have an error and corrective action is needed. In our house this will either be cooling or heating depending on whether the PV is higher or lower than the SP respectively.

Let's imagine the temperature PV in our house is higher than the SP. It is too hot. The air-con is switched on and the temperature drops.

The sensor picks up the lower temperature, feeds that back to the controller, the controller sees that the "temperature error" is not as great because the PV (temperature) has dropped and the air con is turned down a little.

This process is repeated until the house has cooled down to 22°C and there is no error.

Then a disturbance hits the system and the controller has to kick in again.

In our house the disturbance may be the sun beating down on the roof, raising the temperature of the air inside.

So that's a really, really basic overview of a simple feedback control system. Sounds dead simple eh?

Understanding the controller

Unfortunately, in the real world we need a controller that is a bit more complicated than the one described above, if we want top performance form our loops. To understand why, we will be doing some "thought experiments" where we are the controller.

When we have gone through these thought experiments we will appreciate why a PID algorithm is needed and why/how it works to control the process.

We will be using the analogy of changing lanes on a freeway on a windy day. We are the driver, and therefore the controller of the process of changing the car's position.

Here's the Block Diagram we used before, with the labels changed to represent the car-on-windy-freeway control loop.

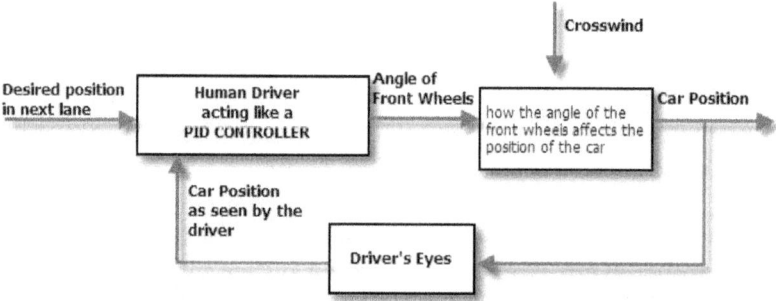

Notice how important closing the feedback loop is. If we removed the feedback loop we would be in "open loop control", and would have to control the car's position with our eyes closed!

Thankfully we are under "Closed loop control" -using our eyes for position feedback.

As we saw in the house-temperature example the controller takes the both the PV and SP signals, which it then puts through a black box to calculate a controller output. That controller output is sent to an actuator which moves to actually control the process.

We are interested here in what the black box actually does, which is that it applies 1, 2 or 3 calculations to the SP and Measured PV signals. These calculations, called the "Modes of Control" include:

- Proportional (P)
- Integral (I)
- Derivative (D)

Under The Hood Of The PID Controller

Here's a simplified block diagram of what the PID controller does:

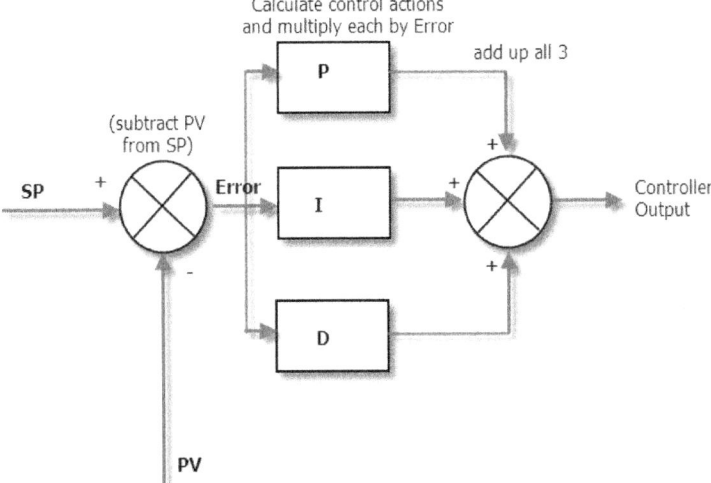

It is really very simple in operation. The PV is subtracted from the SP to create the Error. The error is simply multiplied by one, two or all of the calculated P, I and D actions (depending which ones are turned on). Then the resulting "error x control actions" are added together and sent to the controller output.

These 3 modes are used in different combinations:

P – Sometimes used

PI - Most often used

PID – Sometimes used

PD – rare as hen's teeth but can be useful for controlling servomotors.

Derivatives

Go into the control room of a process plant and ask the operator:

"What's the derivative of reactor 4's pressure?"

And the response will typically be:

"Bugger off smart arse!"

However go in and ask:

"What's the rate of change of reactor 4's pressure?"

And the operator will examine the pressure trend and say something like:

"About 5 PSI every 10 minutes"

He's just performed calculus on the pressure trend! (don't tell him though or he'll want a pay raise)

So derivative is just a mathematical term meaning rate-of-change. That's all there is to it.

Integrals without the Math

Is it any wonder that so many people run scared from the concept of integrals and integration, when this is a typical definition?

Integral

From Wikipedia, the free encyclopedia

This article is about the concept of integrals in calculus. For the set of numbers, see integer. For other uses, see,

Integration is a fundamental concept in mathematics, specifically in the field of calculus and, more broadly, mathematical analysis. Given a function f of a real variable x and an interval $[a,b]$ of the real line, the integral

$$\int_a^b f(x)\,dx,$$

is defined informally to be the signed area of the region in the xy-plane bounded by the graph of f, the x-axis, and the vertical lines $x = a$ and $x = b$.

The term "integral" may also refer to the notion of ant (derivative, a function F whose derivative is the given function f. In this case it is called an indefinite integral, while the integrals discussed in this article are termed definite integrals. Some authors maintain a distinction between antiderivatives and indefinite integrals.

The principles of integration were formulated independently by Isaac Newton and Gottfried Leibniz in the late seventeenth century. Through the fundamental theorem of calculus, which they independently developed, integration is connected with differentiation: if f is a continuous real-valued function defined on a closed interval $[a, b]$, then, once an antiderivative F of f is known, the definite integral of f over that interval is given by

$$\int_a^b f(x)\,dx = F(b) - F(a).$$

What the!?!?

If you understood that you are a smarter person than me.

Here's a plain English definition:

The integral of a signal is the sum of all the instantaneous values that the signal has been, from whenever you started counting until you stop counting.

So if you are to plot your signal on a trend and your signal is sampled every second, and let's say you are measuring temperature. If you were to superimpose the integral of the signal over the first 5 seconds – it would look like this:

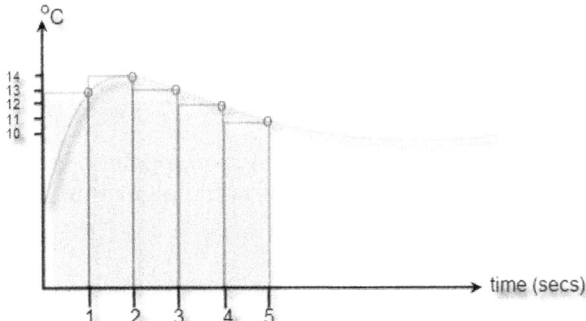

The green line is your temperature, the red circles are where your control system has sampled the temperature and the blue area is the integral of the temperature signal. It is the sum of the 5 temperature values over the time period that you are interested in. In numerical terms it is the sum of the areas of each of the blue rectangles:

$$(13 \times 1)+(14 \times 1)+(13 \times 1)+(12 \times 1)+(11 \times 1) = 63 \, °C \, s$$

The curious units (degrees Celsius x seconds) are because we have to multiply a temperature by a time – but the units aren't important.

As you can probably remember from school –the integral turns out to be the area under the curve. When we have real world systems, we actually get an approximation to the area under the curve, which as you can see from the diagram gets better, the faster we sample.

Proportional Control

Here's a diagram of the controller when we have enabled only P control:

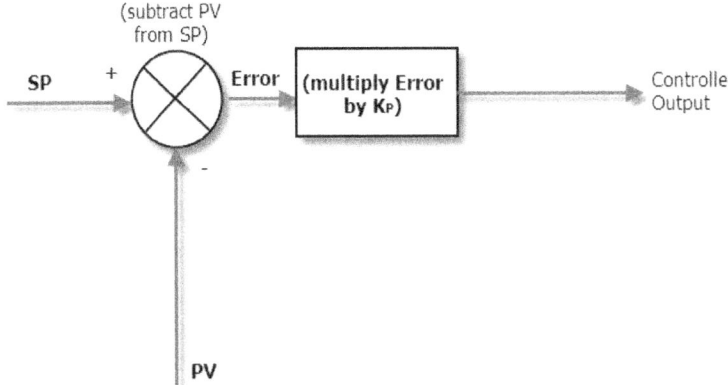

In Proportional Only mode, the controller simply multiplies the Error by the Proportional Gain (K_p) to get the controller output.

The Proportional Gain is the setting that we tune to get our desired performance from a "*P* only" controller.

A Match Made in Heaven: The P + I Controller

If we put Proportional and Integral Action together, we get the humble PI controller. The Diagram below shows how the algorithm in a PI controller is calculated.

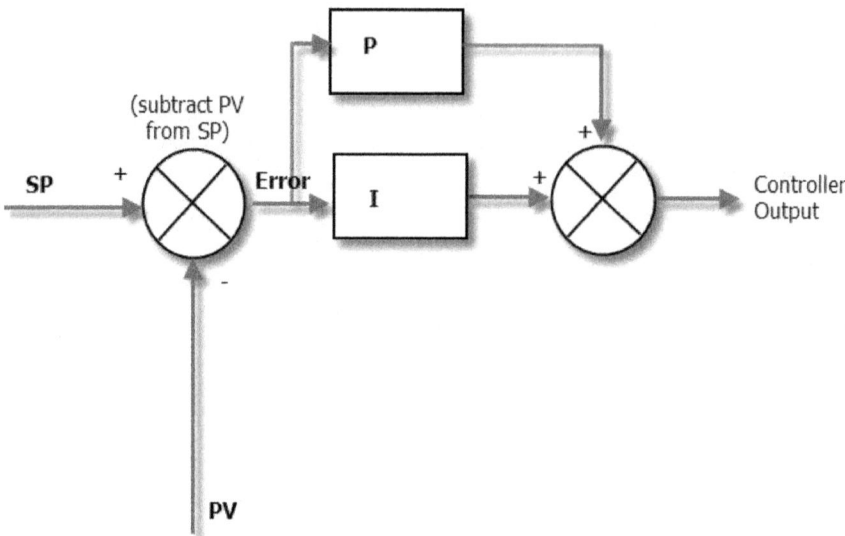

The tricky thing about Integral Action is that it will really screw up your process unless you know **exactly** how much Integral action to apply.

A good PID Tuning technique will calculate exactly how much Integral to apply for your specific process - but how is the Integral Action adjusted in the first place?

Adjusting the Integral Action

The way to adjust how much Integral Action you have is by adjusting a term called "minutes per repeat". Not a very intuitive name is it?

So where does this strange name come from? It is a measure of how long it will take for the Integral Action to match the Proportional Action.

In other words, if the output of the proportional box on the diagram above is 20%, the repeat time is the time it will take for the output of the Integral box to get to 20% too.

And the important point to note is that the "bigger" integral action, the quicker it will get this 20% value. That is, it will take fewer minutes to get there, so the "minutes per repeat" value will be smaller.

In other words the smaller the "minutes per repeat" is the bigger the integral action.

To make things a bit more intuitive, a lot of controllers use an alternative unit of "repeats per minute" which is obviously the inverse of "minutes per repeat".

The nice thing about "repeats per minute" is that the bigger it is - the bigger the resulting Integral action is.

Derivative Action – Predicting the Future

OK, so the combination of P and I action seems to cover all the bases and do a pretty good job of controlling our system. That is the reason that PI controllers are the most prevalent. They do the job well enough and keep things simple. Great.

But engineers being engineers are always looking to tweak performance.

They do this in a PID loop by adding the final ingredient: Derivative Action.

So adding derivative action can allow you to have bigger P and I gains and still keep the loop stable, giving you a faster response and better loop performance.

If you think about it, Derivative action improves the controller action because it predicts what is yet to happen by projecting the current rate of change into the future. This means that it is not using the current measured value, but a future measured value.

The units used for derivative action describe how far into the future you want to look. *i.e.* If derivative action is 20 seconds, the derivative term will project the current rate of change 20 seconds into the future.

The big problem with D control is that if you have noise on your signal (which looks like a bunch of spikes with steep sides) this confuses the hell out of the algorithm. It looks at the slope of the noise-spike and thinks:

"Holy crap! This process is changing quickly, lets pile on the D Action!!!"

And your control output jumps all over the place, messing up your control.

Of course you can try and filter the noise out, but my advice is that, unless PI control is really slow, don't worry about switching D on.

ON-OFF CONTROL

This is the simplest form of control, used by almost all domestic thermostats. When the oven is cooler than the set-point temperature the heater is turned on at maximum power, M, and once the oven is hotter than the set-point temperature the heater is switched off completely. The turn-on and turn-off temperatures are deliberately made to differ by a small amount, known as the hysteresis H, to prevent noise from switching the heater rapidly and unnecessarily when the temperature is near the set-point. The fluctuations in temperature shown on the graph are significantly larger than the hysteresis, as can be confirmed with the interactive simulation, due to the significant heat capacity of the heating element.

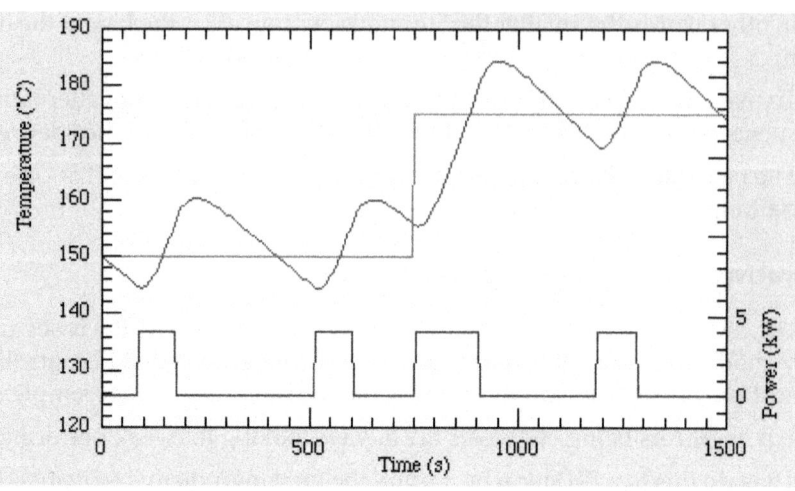

Proportional Control

A proportional controller attempts to perform better than the On-Off type by applying power, W, to the heater in proportion to the difference in temperature between the oven and the set-point,

$$W = P \times (Ts - To)$$

where P is known as the *proportional gain* of the controller. As its gain is increased the system responds faster to changes in set-point but becomes progressively underdamped and eventually unstable. The final oven temperature lies below the set-point for this system because some difference is required to keep the heater supplying power. The heater power must always lie between zero and the maximum M because it can only source, not sink, heat.

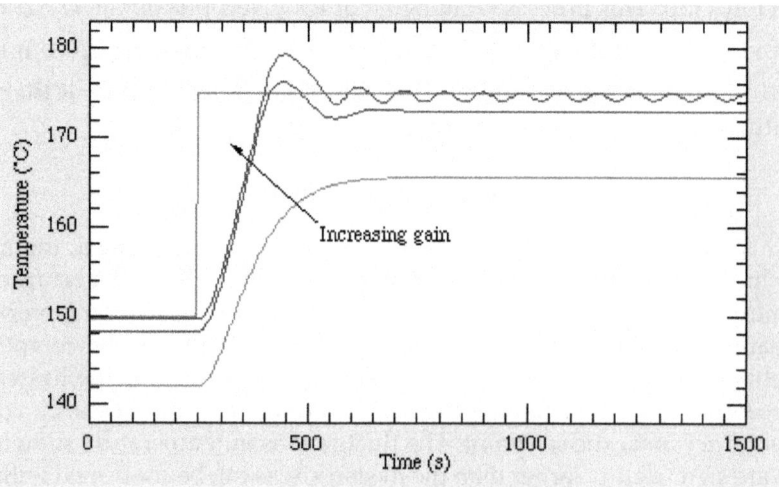

Proportional+Derivative Control

The stability and overshoot problems that arise when a proportional controller is used at high gain can be mitigated by adding a term proportional to the time-derivative of the error signal,

$$W = P \times ((Ts - To) + D \times ddt(Ts - To))$$

This technique is known as *PD control*. The value of the *damping constant, D*, can be adjusted to achieve a critically damped response to changes in the setpoint temperature, as shown in the next figure.

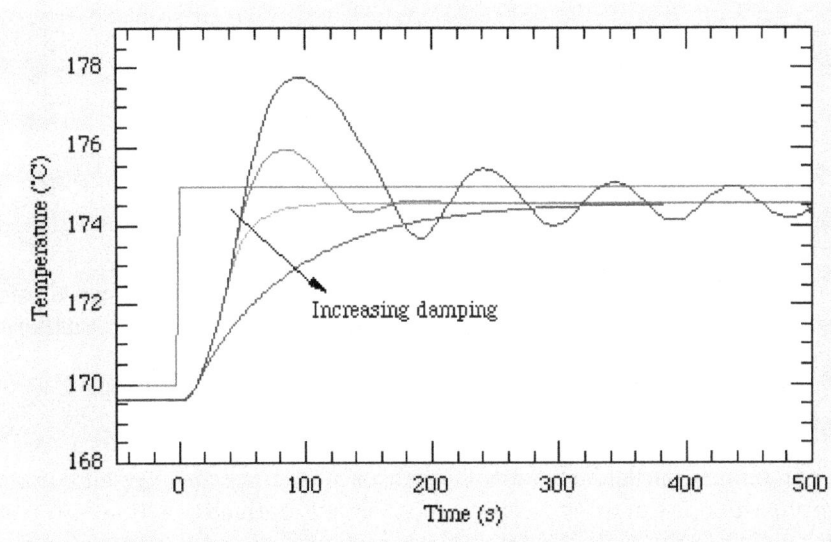

Too little damping results in overshoot and ringing, too much causes an unnecessarily slow response.

Proportional + Integral + Derivative Control

Although PD control deals neatly with the overshoot and ringing problems associated with proportional control it does not cure the problem with the steady-state error. Fortunately it is possible to eliminate this while using relatively low gain by adding an integral term to the control function which becomes

$$W = P \times ((Ts - To) + D \times ddt(Ts - To) + I \times \int(Ts - To)dt)$$

where *I*, the *integral gain* parameter is sometimes known as the controller *reset level*. This form of function is known as proportional-integral-differential, or PID, control. The effect of the integral term is to change the heater power until the time-averaged value of the temperature error is zero. The method works quite well but complicates the mathematical analysis slightly because the system is now third-order.

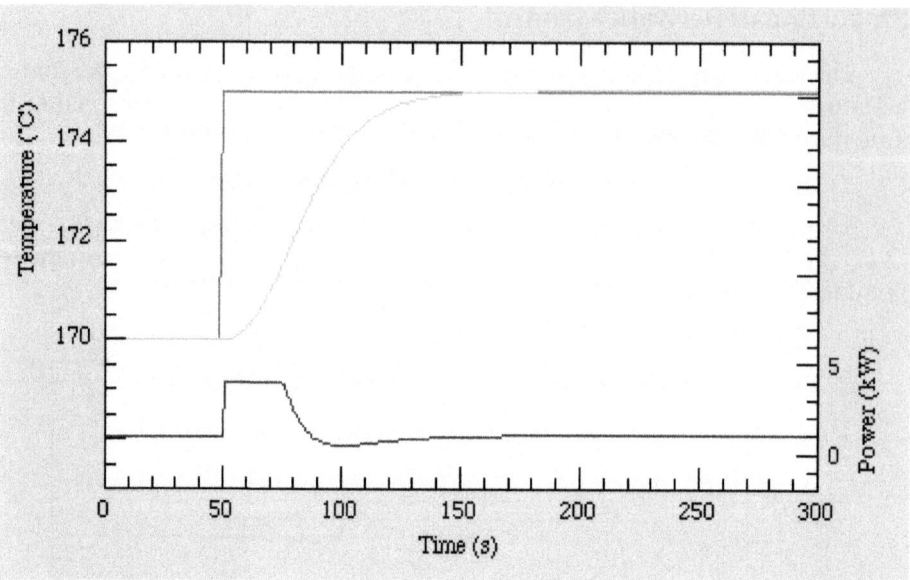

The figure shows that, as expected, adding the integral term has eliminated the steady-state error. The slight undershoot in the power suggests that there may be scope for further tweaking.

Proportional + Integral Control

Sometimes, particularly when the sensor measuring the oven temperature is susceptible to noise or other electrical interference, derivative action can cause the heater power to fluctuate wildly. In these circumstances it is often sensible use a PI controller or set the derivative action of a PID controller to zero.

Third-Order Systems

Systems controlled using an integral action controller are almost always at least third-order. Unlike second-order systems, third-order systems are fairly uncommon in physics but the methods of control theory make the analysis quite straightforward. For instance, applying the so-called *Routh-Hurwitz stability criterion*, which is a systematic way of classifying the complex roots of the auxiliary equation for the model, it can be shown that provided the integral gain is kept sufficiently small then parameter values can be found to give an acceptably damped response with the error temperature eventually tending to zero if the set-point is changed by a step or linear ramp in time. Whereas derivative control improved the system damping, integral control eliminates steady-state error at the expense of stability margin.

PRACTICAL MATTERS

In its raw form integral control can be a mixed blessing; if the error $T_s - T_o$ is large for a long period, for example after a large change in T_s or at switch-on, the value of the integral can become excessively large and cause overshoot or undershoot that takes a long time to recover. To avoid this problem, which is often called 'integral wind-up', sophisticated controllers will inhibit integral action until the system gets fairly close to equilibrium. One method of achieving this is used by the interactive simulation: when the "Limit I?" option is selected the value of the integral is held constant during periods when the heater is at maximum, or zero, power. This technique seems quite effective and would be straightforward to incorporate in a real controller.

Any system using a resistive electrical heater to control temperature is inherently non-linear because an electrical heater can only generate, not absorb, heat. When the oven temperature is higher than the set-point cooling occurs at a rate that depends on the oven and its temperature not the controller and dual PID controllers allow different heating and cooling parameter values to cope with this. It is possible to build your own PID controller from a few operational-amplifiers. Commercial PID process controllers vary in cost between £75 for a simple model and £600 for an intelligent autotuning dual PID model.

Don't just assume that the knobs on a PID controller correspond to the parameters defined in this document. Values are sometimes specified by time constants in which case a long integral time constant is equivalent to a low value of I but a long derivative time constant means a large value of D. The proportional gain is sometimes set by choosing a proportional band which is the change in input that gives maximum change in heater power so a small number for this corresponds to a large value of P.

Varieties of PID Algorithms

The *parallel algorithm* variety of PID control, the version discussed

$$W = P \times ((Ts - To) + D \times ddt(Ts - To) + I \times \int(Ts - To)dt)$$

is often referred to as the 'ideal algorithm'. To implement this scheme accurately one needs at least three amplifiers (the example controller circuit uses five). However, if slight deviations from the 'ideal' behaviour are permitted, only one amplifier is needed. This can be a great advantage, particularly in pneumatic systems where amplifiers are expense items. Differences in the achievable control performance due to which algorithm is being used are not normally significant. However, the tuning procedures used do differ slightly. Also, some controllers only apply derivative action to the process variable, not to the set point. Whether this is an advantage or not depends on the circumstances.

Control Theory

Avoid re-inventing the wheel when tackling difficult feedback or control problems - control theory is a well-developed branch of engineering and has a range of powerful techniques to design and analyse systems involving feedback. As well as having systematic methods for solving complicated problems it introduces the important ideas of *controllability*, *observability* ('Does the system have distinct states that can't be unambiguously identified by the controller?') and *robustness*('Will control be regained satisfactorily after an unexpected disturbance?').

Noise and the Frequency Domain

The frequency domain behaviour of the model can be investigated with the interactive simulator which will plot the open- and closed-loop frequency response for the system. As the controller gain is increased the phase margin reduces towards zero causing the overshoot described previously and a resonant peak in the frequency domain response. Any additional lag in the system, for example a non-negligible time-constant for the sensor measuring T_o, will make it possible for the system to oscillate, which is the reason for the second step in the procedure suggested for tuning a PID controller. Note that integral action reduces the phase margin, derivative action improves it. Even if a system is technically stable it is unwise to operate it with a large peak in the closed-loop gain as this will act as an amplifier for any sensor noise and may cause large and undesired fluctuations in the heater power. If you have a noisy system to control you almost certainly do not want to use any derivative action.

TUNING A PID TEMPERATURE CONTROLLER

In some case one may be able to measure the oven time constants directly and hence calculate the best controller settings. Often an equipment manufacturer will have suggested settings based on their commissioning report - a good reason read the manual first. Sometimes one has no option but to set up, or 'Tune', a system in closed-loop mode by trial and error so here are two straightforward procedures to tune a PID-controlled oven, they will get fairly close to optimum settings in most cases.

CDHW Method

1. Adjust the set-point value, T_s, to a typical value for the envisaged use of the system and turn off the derivative and integral actions by setting their levels to zero. Select a safe value for the maximum power M and increase the proportional gain until the system is just oscillating.

2. Note the period of oscillation then reduce the gain by 30%.

3. Suddenly decreasing or increasing T_s by about 5% should induce under-damped oscillations. Try several values of derivative level and choose a value for that gives a critically damped response. If the controller is calibrated D will need to be approximately one third of the oscillation period noted above.

4. Slowly increase the integral level until oscillation just starts, then reduce this level by a factor of two or three - this should be enough to stop the oscillation. I have found it is a good idea to use the lowest integral level that gives adequate performance.

5. Check the overall performance of system is satisfactory under the conditions it will be used.

This procedure is based on the assumption that a critically damped system is optimal and the fact that stability and noise must be traded for response time. Please bear in mind that the second step may involve large temperature oscillations and so the procedure would not be suitable if these could be dangerous or cause damage, for example in a chemical processing plant.

John Shaw's (Ziegler-Nichols Based) Method

1. Adjust the set-point value, T_s, to a typical value for the envisaged use of the system and turn off the derivative and integral actions by setting their levels to zero. Select a safe value for the maximum power M and set the proportional gain to minimum.

2. Progressively increase the gain until suddenly decreasing or increasing T_s by about 5% induces oscillations that are just self-sustaining.

3. The gain at this stage will be set to the ultimate gain G_u the period of the oscillations is known as the ultimate period t_u. Note the values of each quantity.

4. Set the controller parameters as follows:

   o   P-Control: $P=0.50*G_u$, $I=0$, $D=0$.

   o   PI-Control: $P=0.45*G_u$, $I=1.2/t_u$, $D=0$.

   o   PID-Control: $P=0.60*G_u$, $I=2/t_u$, $D=t_u/8$.

5. Check the overall performance of system is satisfactory under the conditions it will be used.

This procedure was adapted slightly from John Shaw's, description of the Ziegler-Nichols Closed Loop method. It should yield a system that is slightly underdamped; if a less "aggressive" response is desired try reducing P to half the values listed. As was the case with the CDHW method the second step may involve large temperature oscillations and so the procedure would not be suitable if these could be dangerous or cause damage, for example in a nuclear reactor. Strictly speaking, the Ziegler-Nichols method was developed for the traditional *series*, or *interacting* design of controller.

CONTROLLER CIRCUIT

This circuit is the basis of a temperature controller. Study it and then answer the questions that follow. The questions have links to outline answers but please resist the temptation to look at these until you have written down your own answers to all the questions.

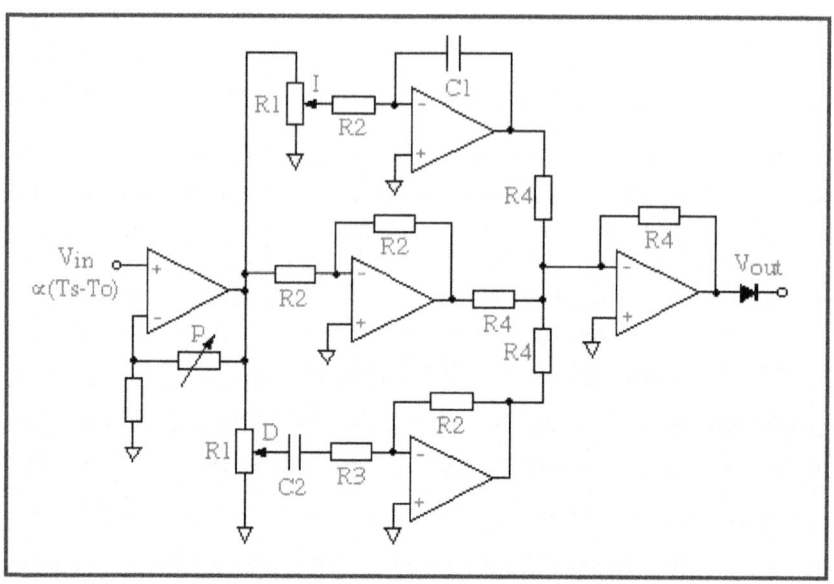

TEMPERATURE CONTROLLER SIMULATION

Instructions

The parameters are described elsewhere, with links to explanations and definitions in the main document. Their values in the boxes can be edited in the usual way. With the present values the set-point temperature T_s will stay at 150 °C until 100 s into the simulation when it will ramp linearly to 175 °C at 200 s. Either On-Off or PID controllers can be selected with the first menu which also has an option "PID Bode" to display the frequency-domain response of the PID system. The second menu changes the temperature units.

On the time-domain displays the red line is the set-point temperature, T_s, the green line is the oven temperature, T_o, and the blue line is the heater power, W. On the frequency-domain displays the red curve is the open-loop gain of the system, and the black curve is its phase. The blue curve is the response of the oven temperature T_o to changes in heater power W and the green curve is the closed-loop response of the oven temperature to changes in the set-point temperature T_s. There is a more technical discussion of aspects of this simulation available.

Model Parameters		Simulation Parameters		Controller Parameters
$T_e =$ 25	°C ▼	ON-OFF ▼		$H =$ 10 °C
$R_o =$ 0.1	°C/W	Run for 1000 s		$P =$ 900 W/°C
$C_o =$ ··	J/°C	$T_s =$ 150 °C until	100 s	$I =$ 2.0e-4 /s
$C_h =$ 500	J/°C	$T_s =$ 175 °C after	200 s.	$D =$ 3 s
$R_{ho} =$ 0.15	°C/W			$M =$ 4000 W
Sensor lag= 2.5	s	Noise= 0.0	°C/sqrt(Hz)	Limit I?

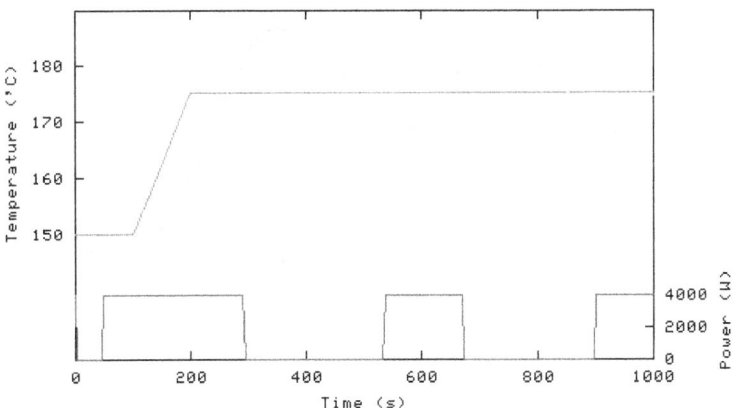

Chapter 11

Hardware Implementation of Multiple Fan Beam Projection Technique in Optical Fibre Process Tomography

Ruzairi Abdul Rahim[1,*], Mohd Hafiz Fazalul Rahiman[2], Leong Lai Chen[1], Chan Kok San[1] and Pang Jon Fea[1]

[1]   Control & Instrumentation Engineering Department, Faculty of Electrical Engineering, Universiti Teknologi Malaysia, 81310 Skudai, Johor Bahru, Malaysia
[2]   School of Mechatronic Engineering, Universiti Malaysia Perlis, 02600 Arau, Perlis, Malaysia; E-Mail: hafiz@unimap.edu.my

*   Author to whom correspondence should be addressed; E-mail: ruzairi@fke.utm.my

ABSTRACT

The main objective of this project is to implement the multiple fan beam projection technique using optical fibre sensors with the aim to achieve a high data acquisition rate. Multiple fan beam projection technique here is defined as allowing more than one emitter to transmit light at the same time using the switch-mode fan beam method. For the thirty-two pairs of sensors used, the 2-projection technique and 4-projection technique are being investigated. Sixteen sets of projections will complete one frame of light emission for the 2-projection technique while eight sets of projection will complete one frame of light emission for the 4-projection technique. In order to facilitate data acquisition process, PIC microcontroller and the sample and hold circuit are being used. This paper summarizes the hardware configuration and design for this project.

Keywords

Optical tomography, Fan Beam, projection, optical fibres.

1. PROCESS TOMOGRAPHY OVERVIEW

The widespread need for direct analysis of the internal characteristics of process plants in order to improve the design and operation of equipment has made process tomography a main research activity within the industrial instrumentation. Originated from the Greek words *'tomos'* which means slice and *'graph'* meaning picture, tomography can be defined as a picture of a slice [1]. In simple terms, tomography is an imaging technique that enables one to determine the contents of a closed system without physically looking inside it.

There are different requirements in an industrial environment than there are within a medical one: different regulations regarding for example use of ionising modalities and different speed requirements [2] Technically, Process Tomography can be described as imaging process parameters in space and time. Important flow information such as concentration measurement, velocity, flow rate, flow compositions and others can be obtained without the need to invade the process or object. As a result, cross sectional images of processes generate better online inspection, monitoring and process control -promoting improved yields and more effective utilization of available process capacity. Potentially, tomographic systems may also be an alternative approach in developing and verifying process theories and models, as well as for improving process instrumentation.

The earlier researches done by Ruzairi [3], Sallehuddin [4], Khoo [5] and Hisyamuddin [6] have shown that the optical fibre sensor is applicable in flow visualization (image reconstruction). The acquired concentration profile from the image reconstruction is needed together with the velocity profile to complete the mass flow rate estimation in a pneumatic conveying system. Basically, the principle of measurements in tomography is to obtain all possible combinations of measurements from the sensor system. The higher the measurements obtained from the sensors, the resolution of the system would be better.

By using the parallel projection, previous researches have each faced the problem of obtaining a high resolution of their system. This is because the parallel projection method limits the number of measurements to the number of sensors being used. In his research, Chan [7] has implemented the switch-mode fan beam projection technique to obtain flow visualization using LED as light source but resolution and the number of sensors in his system is limited by the physical size of the LED emitters. Thus, this research will focus in implementing the multiple fan beam method using optical fibre sensors to increase both the number of sensors and number of measurements in order to obtain a system with high resolution.

2. INTRODUCTION TO THE HARDWARE SYSTEM

A typical Optical Fibre Process Tomography (OFPT) system consists of the sensor's array, signal control and conditioning circuit, data acquisition system (DAS) and also the display unit, namely the computer. Topology of the constructed hardware system in this research is illustrated in Figure 1.

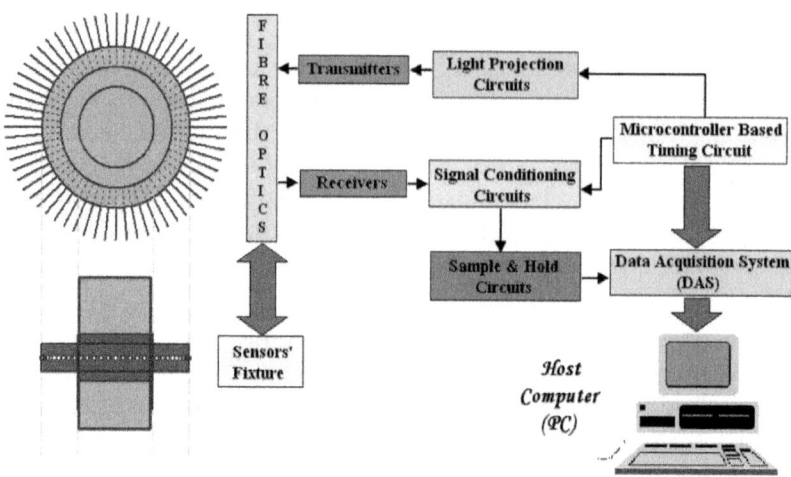

Figure 1. Topology of the hardware construction.

By referring to Figure 1, the micro controller is used to control the duration of light projection, sample and hold digital input and data acquisition system (DAS) synchronization signals. Through the fibre optics, photodiodes detect the physical signals (light beams) from the transmitters. In the signal conditioning circuit, the physical signals are being converted into voltage readings and then amplified. The analogue signals go through the sample and hold circuit before being transferred into DAS. DAS then converts the analogue signals into digital signals. These digital signals are being sent to the PC for image reconstruction.

By referring to Figure 1, the micro controller is used to control the duration of light projection, sample and hold digital input and data acquisition system (DAS) synchronization signals. Through the fibre optics, photodiodes detect the physical signals (light beams) from the transmitters. In the signal conditioning circuit, the physical signals are being converted into voltage readings and then amplified. The analogue signals go through the sample and hold circuit before being transferred into DAS. DAS then converts the analogue signals into digital signals. These digital signals are being sent to the PC for image reconstruction.

3. SELECTION OF EMITTERS AND RECEIVERS

Emitters and receivers are the main optical sensors that must be carefully chosen to satisfy the characteristics and requirements of the hardware system. According to Chan [7], for a system which implements the switch-mode fan beam projection, the emitter must have a very fast setting time when driven by a pulse current while the receiver must have fast transient characteristic when exposed to the switched light sources. The selection of the sensors is based on the specified requirements such as size of the emitting area, angular spread of the emitted light, reliability, physical size, dynamic response and costs of the light source.

For the emitters, three types of optical devices are being considered and they are the light emitting diode (LED), infrared (IR) and laser diodes. Although the laser diodes have a fast operational speed, the LED is generally user friendly and is certainly more cost effective when compared with laser diodes. Besides that, the output power of the LED is linearly proportional to driving the current while laser diodes have an output power which is proportional to current above the threshold.

Linearity is an important characteristic to light sources in analog applications which is emphasized in the implementation of the OFPT sensors. Based on the comparisons of the LED and laser diodes in terms of linearity, costs and it is found that the LED is a better choice of emitter for this project.

However, there is a weakness in LED to be used as transmitter because the light of LED is visible light with the wavelength in between 380-700 nm and therefore results in the tomography sensor designed is easily getting noise from the surrounding environment light source [8]. Most lights sources that we use daily are white or visible lights such as the incandescent lamp (light bulb) and fluorescent light which have a peak of radiant power at 550 nm that can simply affect the light received by the photo-receivers.

The most suitable part of the spectrum of light which is suitable to be selected as light source for this project is the infrared. Generally, the wavelength of the infrared LED lies in between 700nm to 1100nm, thus potentially can safeguard the tomography sensor from being affected by visible light.

The selected transmitter used is the SFH484-2 GaAIA infrared emitter with its peak wavelength at 880n. The small radiation angle is necessary because the emitting area needed for the infrared to be coupled with the fibre optics is small and narrow.

Meanwhile, the main requirements to choose the photo-receivers is to select a photo-receiver with high sensitivity, fast switching time (taking into account the transient/rise and fall time), cost effective and most compatible with the selected infrared emitter. Phototransistors generally have a slower response than photodiodes [9] and linearity of the phototransistor is over a much narrower range than a photodiode. Rise time of the phototransistor is poor due to the combined capacitance of the B/E and C/E junctions and the lifetimes of the carriers in the depletion region of the junction. Based on the need of a fast response and high sensitivity, the photodiode is selected as the photo-sensor for this hardware system instead of the phototransistor.

Basically, all the photodiode models have a fast switching time of 5ns and they have the same diameter. Thus, selection is mainly based on the price and also the spectral range. The first model, SFH203P is apparently the cheapest but a main concern is that if has a very wide spectral range from 400nm to 110nm. As stated earlier, the visible light has a range of 380nm to 700nm, thus this photodiode performance might be influenced by the visible light from the environment. The second model, SFH203PFA has a narrow spectral range but the price is too expensive. The most reasonably priced and has the most agreeable spectral range is the SFH213-FA photodiode and this will be selected to match with the infrared transmitter. Besides that, the SFH213-FA has a fast transient time which can reduce the signal setting time [10].

3.1 Preparation of Optical Fibres

In the area of tomographic imaging, an initial investigation into using fibres optic as measurement sensors in pneumatic conveying was started in Sheffield Hallam University [3]. In using optical fibre for tomography imaging, the basic optical transmitter converts electrical input signals into modulated light for transmission over an optical fibre. Also, the light beam from the transmitter is being received by the receiver via the fibre optic. This configuration is being illustrated in Figure 2.

With regard to its small physical size, it is believed that using fibre optic will allow a higher number of optical sensors to be installed, thus achieving high-resolution measurement in optical tomography. It is also said that the optical fibre sensors provide wide bandwidth which enables measurements to be performed on high speed flowing particles [4]. As stated earlier in this paper, the optical fibres are used together with the selected transmitters and receivers. The choice of using single core polymer cable fibre optic (with core diameter at 1.00mm and overall diameter at 2.25mm) instead of the fibres made of glass is because the former is more affordable, easier to install [11] and since the core is made of plastic instead of glass, terminating the cable will be easier [3]. Figure 3 shows the fibre optic after treatment.

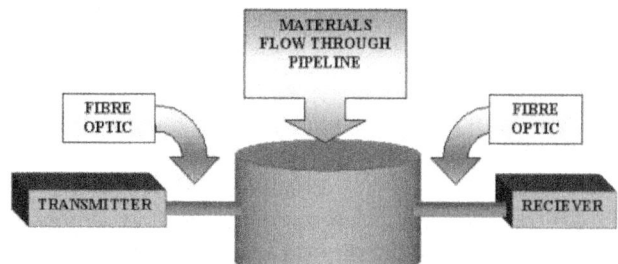

Figure 2. Fibre optics configurations.

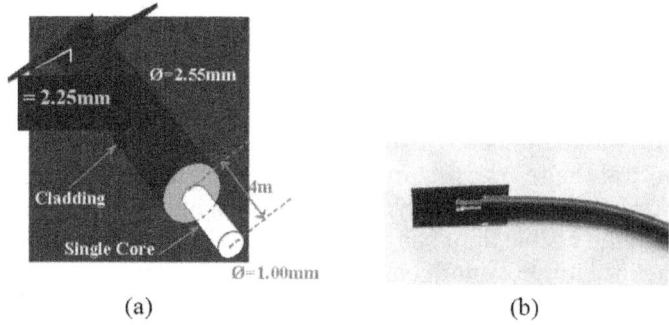

(a)   (b)

Figure 3. (a) Optical fibre end termination **(b)** Drawing Actual photograph.

The fibre optic has a numerical aperture of 0.47 and acceptance angle of 28 degrees as stated in the data sheet. The numerical aperture determines the acceptance cone of the fibre [12]. Equation1 gives us the formula to calculate the numerical aperture and Figure 4 shows the acceptance angle of an optical fibre. The total receiving angle for the fibre optic is two times the acceptance angle and in this case, it is 56 degrees.

$$NA = \sin \theta_A \tag{1}$$

Whereby:

NA = numerical aperture of the fibre optic.

θ_A = acceptance angle of the fibre optic.

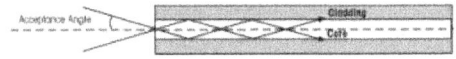

Figure 4. The acceptance angle for optical fibre.

Unlike in the application of optical fibre sensors in parallel beam projection, the emission beam should not concentrate in a straight line. Instead, the emitted fan beam should have a transmission angle. Preliminary testing shows that the maximum achievable emission angle for the fibre optic transmitter is about 30°, after the fibre optic is being lensed. There are 32 fibre optic transmitters that are being used in this research; thus in order to make sure that the emission angle is approximately the same, each of the fibre optic emission angles is being tested experimentally as illustrated in Figure 5.

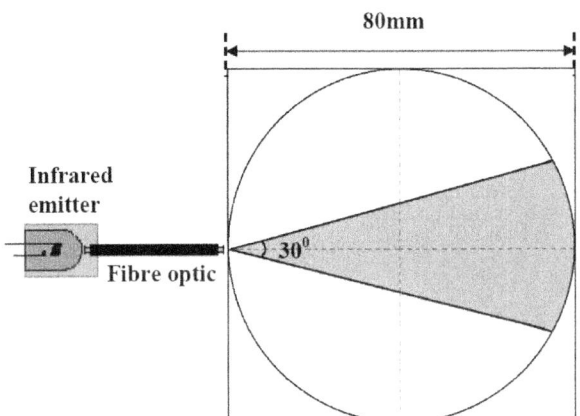

Figure 5. Determining emission angle.

3.2 Fibre optic coupling

Signal loss due to improper coupling between sensors and fibre optics can cause inaccurate data acquisition. In order to avoid transmission loss when the fibre optics is coupled with the infrared emitters and photodiodes, custom-made housing is being used. The housing is made of PVC and designed as such to hold both the infrared emitters and photodiodes with fibre optics to make sure that the connection area is small and the lights can be directed straight, either from the emitters to the fibre optics or light from the fibre optics to the receivers. Figure 6 shows the coupling between the sensor and fibre optic while figure 7 shows the actual photographs of the fibre optic housing and its coupling with the sensor and fibre optic.

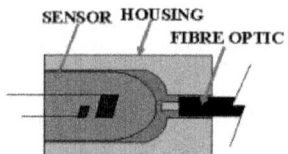

Figure 6. Coupling between sensor and fibre optic.

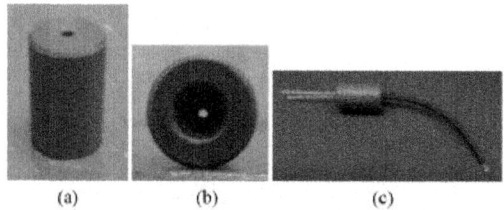

Figure 7. PVC housing **(a)** 3-dimensional view **(b)** Bottom view **(c)** Coupling with infrared emitter and fibre opti.

3.3 Optical Fibre Sensor Fixture Design

A custom-made sensor fixture is being designed to hold and support the optical fibres. The sensor fixture is also made of PVC and 64 holes (each with a diameter of 2. 3mm) are being drilled along the periphery of the fixture. For a pipe

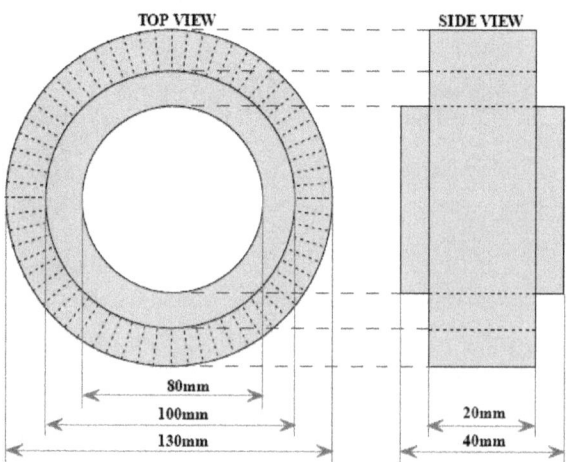

Figure 8. Top view and side view of the sensor fixture.

with 80mm inner diameter, the diameter of the fixture peripheral that supports the fibre optics is 100mm as shown in Figure 8. Figure 9 further shows the actual photography of the optical fibre sensor fixture.

(a)   (b)

Figure 9. Optical fibre sensor fixture: **(a)** 3-dimensional view **(b)** Internal view

4. THE SIGNAL PROCESSING CIRCUITS

4.1 Infrared Projection Circuit

Infrared projection circuit to supply current for infrared transmitters to transmit light. The infrared projection circuit used in this research is shown in Figure 10 below using the basic components of 1Ω resistor and ZTX1048A transistor.

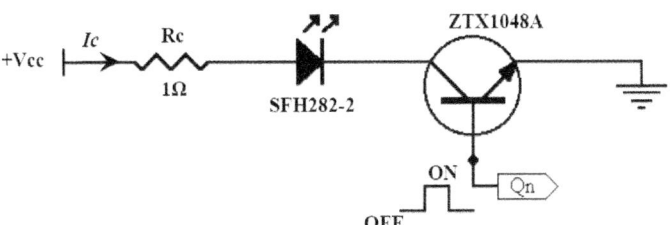

Figure 10. Infrared projection circuit.

From the above circuit, the collector current, I_c can be calculated using Equation 2

$$I_c = \frac{V_{cc} - V_f - V_{CE(sat)}}{R_c}$$

(2)

Whereby:

I_c = collector current or forward current for the projection circuit.

V_{cc} = voltage supply which is 4.5V in this circuit.

V_f = forward voltage of the SFH484-2 which is 3V as stated in datasheet.

$V_{CE(sat)}$ = collector-emitter saturation voltage for the ZTX1048A transistor, which is 245mV as stated in datasheet.

R_c = resistor with the value of 1Ω.

Thus, with the given values, I_c can be calculated as follows:

$$I_c = \frac{5 - 3 - 0.245}{1} = 1.755\,A$$

From Figure 10, the point Q_n at the base of the transistor is connected to the decoder in the signal control circuit. The n represents the n-th decoder output pin as there are 32 individual infrared projection circuits. Base current determines the 'on' and 'off' state of the projection circuit. Positive pulse that is supplied will activate the projection circuit while negative pulse will deactivate the circuit. This operation mode is known as the pulsed mode. The emitter is operated in pulsed mode because it can handle a larger current and hence generate a greater intensity of radiation [7]. For example, for a certain infrared, if applied with continuous current, the maximum achievable forward current is 100mA while pulsed current might be able to withstand forward current up to 1A for 5ms [13] .

4.2 Signal Conditioning Circuit

Signal conditioning circuit which functions as the current to voltage converter for the physical signals received by photodiodes. The voltage readings are further amplified to an acceptable level to be observed. The principle of an optical tomography system is to investigate the light attenuation level for each detector. The signal conditioning circuit for this hardware system is divided into two stages which are the current-to-voltage converter stage and the voltage amplification stage. Figure 11 illustrates the current-to-voltage converter circuit that is used.

The current output response of the photodiode is linearly related to the incident light energy. A monitor of this current must have zero input impedance to response with no voltage across the photodiode. Zero impedance is the role of an op-amp virtual ground as high-amplifier loop gain removes voltage swing from the input [14]. In another words, main job of the op-amp is to adjust the output such that the inverting input equals the non-inverting input. This is the key to the basic current-to-voltage connection circuit as shown in Figure 11.

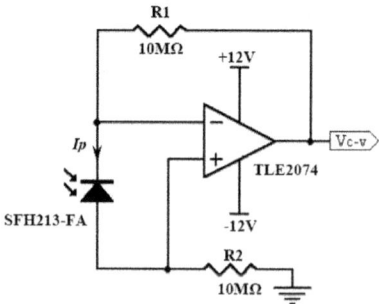

Figure 11. Current-to-voltage converter circuit.

The output voltage of the pre-amp, Vc-v is obtained by using Equation 3.

$$V_{c-v} = I_p (R_1) \tag{3}$$

Whereby:

V_{c-v} = pre-amp output voltage.

I_p = photodiode current.

R_1= feedback resistor of 10Ω.

According to Wong [15], the feedback resistor R_1 determines the transimpedance, and hence the sensitivity of the amplifier. Large R_1 increases sensitivity but at the same time reduces the amplifier's bandwidth since it contributes to the pre-amp's input load impedance.

From the first stage of current-to-voltage converter, the pre-amp voltage will be sent to another op-amp to be amplified. Figure 12 shows the non-inverting voltage amplifier circuit [16].

The same TLE2074 op-amp is being used here in this amplifying circuit. The voltage at the inverting input V_n is defined by Equation 4.

$$V_n = \frac{V_o(R_b)}{R_f + R_b}$$

(4)

Whereby:

V_n = voltage at the inverting input.

V_o = output voltage after the amplifying stage.

R_f = variable feedback resistor of 500kΩ.

R_b = resistor of 100kΩ

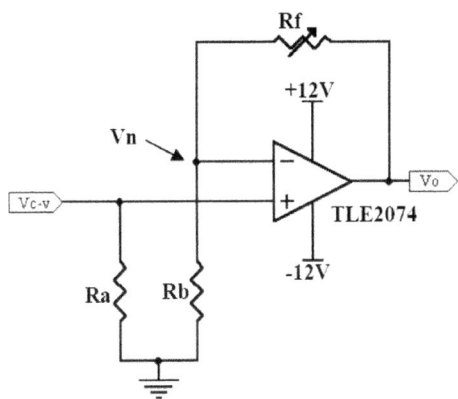

Figure 12. Non-inverting voltage amplifier circuit.

Since the differential voltage is zero, $V_n = V_{c-v}$ and thus, the output voltage can be obtained as referred to Equation 5 (Boylestad and Nashelsky, 1999), with V_o, V_{c-v}, R_f and R_b parameters are as defined earlier.

$$V_o = V_{c-v}\left(1 + \frac{R_f}{R_b}\right)$$

(5)

In this case, the parallel combination values of R_f and R_b results in an approximation of 83.3k Ω. Since R_f has a variable resistive value, R_a is selected to be 100kΩ. This additional resistor R_a is desirable because the voltage drops due to bias current to the inputs are equal and cancel out even over temperature [16]. Thus, the overall performance of the circuit is much improved.

4.3 Microcontroller Signal Controlling Circuit

Previous researchers done by Ruzairi [3], Sallehuddin[4], Chan [7] and Pang [17] used the digital signal control circuit whereby each alteration to the signals needs reconstruction of the logic circuits or devices. This is found to be very troublesome and not flexible when alterations are done to the signal controls.

The main motivation to use the PIC16F84A microcontroller is because the device has sufficient requirements to support the needs of this project. In this signal controlling circuit that has been designed, the micro controller is used to control the duration of light projection, sample and hold digital input and data acquisition system (DAS) synchronization signals. The circuit connection of the microcontroller is shown in Figure 13.

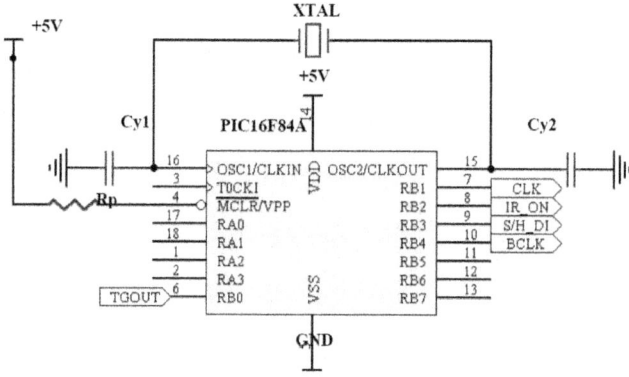

Figure 13. PIC16F84 circuit connection.

With reference to the circuit connection, $\overline{MCLR}$ is the master clear or reset input and has an active low reset to the device. Usually, it is being connected via a resistor to the positive supply pole to prevent from bringing a logical zero to the $\overline{MCLR}$ pin accidentally. This resistor, whose value is selected as 4.7kΩ and its function is to keep a certain line on a logical one as a preventive, is called a pull up. Meanwhile, the XTAL is the crystal oscillator and the value of this crystal used is 2MHz with two ceramic capacitors (a value of 22pF each). For the I/O pins, RB0 functions as an input pin (TGOUT input from the DAS) while RB1, RB2, RB3 and RB4 are output pins for clock (CLK), signal to control the duration of 'on' and 'off' state of the infrareds via decoder (IR_ON), digital input control of sample and hold circuit (S/H_DI) and also the burst clock to signal DAS to start its data conversion process (BCLK).

CLK is the heartbeat to the signal control circuit and is connected to the 74HC161 binary counter while IR_ON will determine the duration of 74HC154 decoder to activate its output to control the light projection circuit. Since the outputs of the 74HC154 have active low outputs [16], the 74HC04 Inverters is being used to toggle the decoder outputs from low to high before connecting to the base of transistors Q_n in the infrared projection circuits. The 74HC04 has six independent inverters [18] and since there are sixteen outputs for the decoder, a total of three inverter chips are needed. The basic connection circuit for the binary counter and the decoder is exemplified in Figure 14.

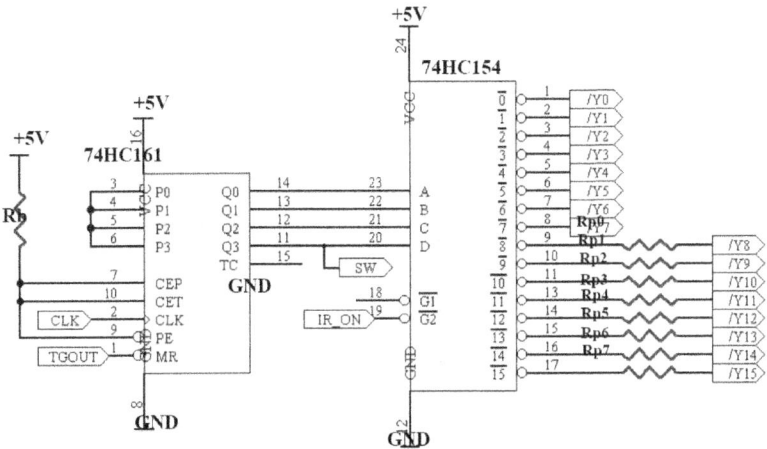

Figure 14. Binary counter and decoder circuit.

A decoder is a logic circuit that accepts a set of input that represents a binary number and activates only the output that corresponds to that input number [19]. The binary input of the decoder is controlled by the high speed, 4-bit 74HC161 binary counter and the counter is activated by the TGOUT input to $\overline{MR}$ which is the master reset for the 74HC161 [20]. The counter will stay idle unless there is a positive edge-trigger which activates it. Once activated, the CLK signal from microcontroller will drive the counter at the programmed frequency.

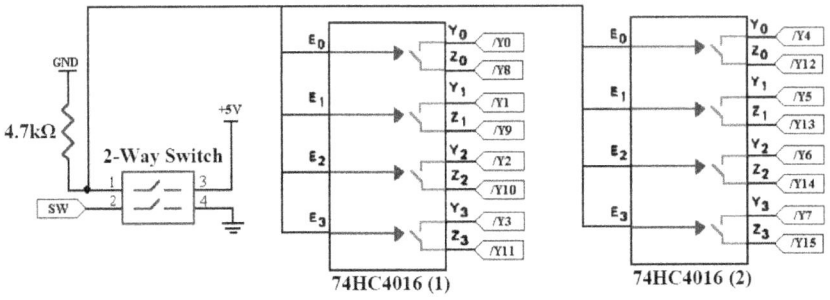

Figure 15. Bilateral and 2-way switch connections.

Not only the outputs of the decoder are connected to the 74HC04 inverters, they are also connected to the 74HC4016 bilateral switch. This bilateral switch is to change the hardware configuration to perform either in the 2-projection or 4-projection mode. Figure 15 shows the connections for the bilateral switches which are connected also to the 2-way switches.

The 74HC4016 has four independent analog switches [21] and has an input control to active the switch. Here, these input controls for all the bilateral switches are connected to the 2-way switch. In the 2-projection mode, both the 2-way switches are left 'open.' In this switch configuration, the outputs of the decoder perform as individual pins. For example, activating /Y0 will set Tx0 and Tx16 to 'on' and /Y8 will set Tx8 and Tx24 to 'on'. If the 2-way switch is in 'closed' mode, current will flow from pin 3 to pin 1 before heading for ground. This way, all the input controls for the bilateral switches will be activated. When this happens, /Y0 will be in the same configuration as /Y8, and therefore, Tx0, Tx16, Tx8 and Tx24 will 'on' to satisfy the 4-projection mode. For the SW input from 74HC161 binary counter, it is referred to ground when used in 4-projection mode. This input represents the MSB which should be connected to ground because the 4-projection mode need only 3bit binary counter.

4.4 Sample and Hold (S/H) Circuit

The sample and hold or S/H function is one which is basic to the data acquisition and A/D conversion process. In most applications, the sample and hold is used as the "front-end" to an A/D converter in data acquisition systems [22]. In these applications, the S/H amplifier is used to store analog data which is then digitized by a relatively slow A/D converter. In this fashion, high speed or multiplexed analog data can be digitized without resorting to complex and expensive ultra-high speed A/D converters [23-24].

Basically, a sample and hold amplifier circuit has two basic and distinct operational states. In the 'SAMPLE' stage, an input signal is sampled and simultaneously transmitted to the output. For the 'HOLD' stage, the last value sampled is held until the input is sampled again. When the S/H goes into the 'HOLD' stage, the S/H switch opens and the voltage stored by the hold capacitor settles through the output buffer. The positive or negative bias current of the output buffer starts charging or discharging the hold capacitor. This degradation of the hold capacitor's voltage over time is known as the "droop rate" [25].

The choice of hold capacitor is important as droop rate is part the major trade-offs in the selection of a hold capacitor value. The leakage of electrolytic and the transient behavior of ceramics rule them out completely in this application. The best choice is probably polypropylene, and after that polystyrene or Mylar [23]. Everything necessary for the S/H except the hold capacitor can be put on chip, so monolithic sample-and-hold circuits, like the LF398, are available and very easy to use. The S/H command is given through a digital logic level, so these circuits interface directly with logic. Besides that, the LF398 has a hold step of less than 1mV, has an acquisition time of 4µs, features high input resistance and also has a low output resistance. Based on these advantages, the LF398 is selected for this research. The S/H circuit is illustrated in Figure 16.

From Figure 16, the Vo is the analog output voltage after amplification from the signal conditioning circuit, while the SSH_DI is the digital logic signal generated by the microcontroller. C_h is the hold capacitor which has a value of 1.5nF and Vout is the output voltage for the sample and hold chip.

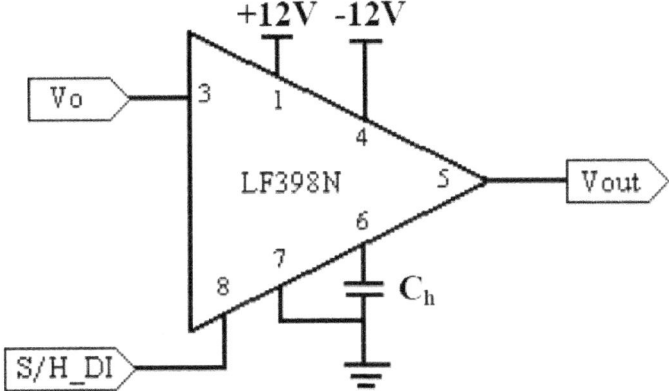

Figure 16. Sample and hold circuit.

5. DATA ACQUISITION PROCESS

For the purpose of converting the analogue signals from the signal conditioning circuit before the data is being processed by the computer for image reconstruction, the Keithley DAS-1802HC high speed data acquisition board has been selected. Figure 17 shows the data acquisition process system.

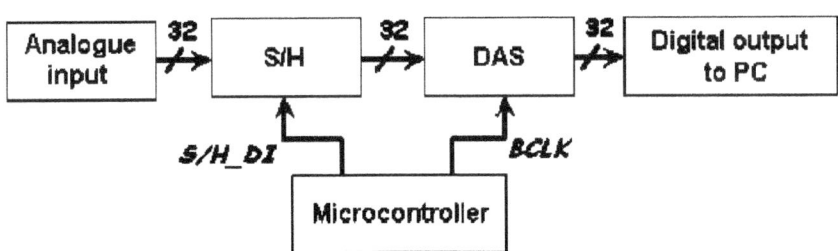

Figure 17. Data acquisition process.

Analogue input from the hardware system goes through the sample and hold circuits before being sent to DAS for analogue to digital conversion. The S/H_DI sends a signal from the microcontroller to the S/H circuit to sample all output signals for a short period of 10µs and then continue to hold the sampled output signals until it receives the next rising edge. At the same time when the S/H signals are on hold, the BCLK signal will send a positive edge signal to the DAS to start data conversion as shown in Figure 18. The total duration of the data conversion time depends on the maximum burst mode clock frequency of 333 kHz in for this DAS.

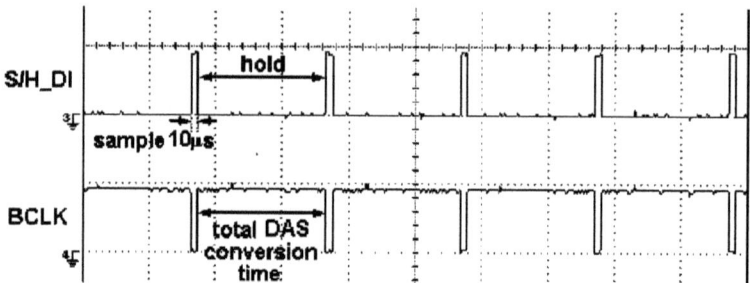

Figure 18. S/H_DI and BCLK signals

When there are many analogue inputs that are needed to be converted into digital outputs, the sample and hold circuits come in handy. For example, in this paper, there are 32 analogue inputs fed in parallel into the 32 channels DAS buffers. A single digital input control signal from the microcontroller will request all 32 individual sets S/H circuits to sample all the analogue signals synchronously. All the signals on-hold will be sent also in parallel to the DAS for data conversion. This will save execution time whereby all 32 analogue signals need not wait to be sampled in serial, which is sampling the 1st, followed by 2nd signal, 3rd signal until 32nd analogue signals. Figure 19 illustrates an example of the analogue and digitalized S/H output signals for Channel 23.

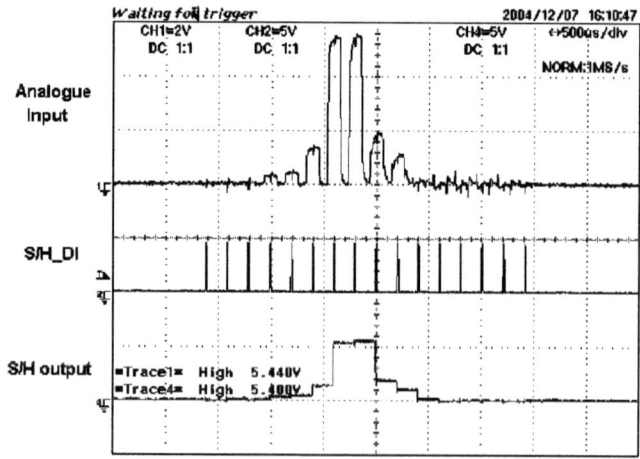

Figure 19. Sample and hold execution.

Figure 20. Actual photographs of the hardware system

6. RESULTS & DISCUSSIONS

6.1 Measured Signals from Oscilloscope

The Yokogawa DL1540 4-Channel Digital Oscilloscope and Tektronik TDS3014 4-Channel Digital Oscilloscope are being used to visualize and also measure the desired waveforms or signals obtained from the hardware. Preliminary results of the hardware development, such as the response of the photo-sensors, microcontroller controlling signals, pre-amp voltages and output voltages will be presented.

6.1.1 Photo-sensors

The various selections of photo-sensors have been discussed previously. Among the topics of discussion is the comparison of the phototransistor and photodiode's performance test in order to select the most suitable photo-sensor. It has been agreed that the SFH213-FA photodiode has been chosen since it is cost effective, has a fast transient time and its spectral range is compatible with the SFH484-2 infrared emitter. However, before the research opted for photodiode as receiver, a few tests are done to proof that the phototransistor has a slower transient time when compared to the phototransistors. For comparison purposes, the BPW85B phototransistor and SFH203-FA photodiode are exposed to a pulsed light (of 5 kHz) from the SFH484-2 infrared emitter. The responses of the photo-receivers are being illustrated in Figure 21.

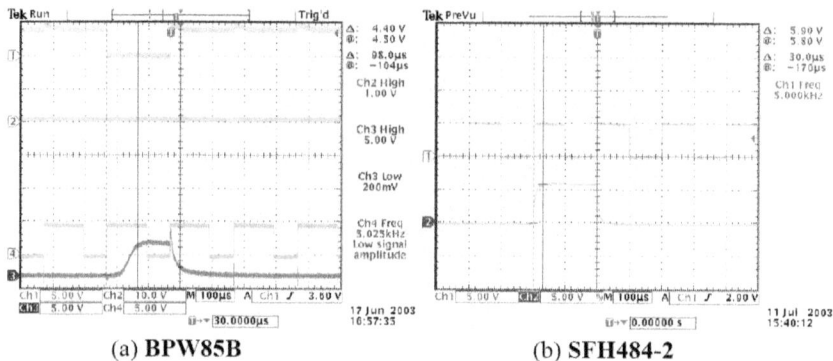

(a) **BPW85B**   (b) **SFH484-2**

Figure 21. Photo-receivers' transient response.

Obviously, the SFH484-2 photodiode (with a transient time of approximately 30μs and fall time also about 30μs) has a faster switching time than the BPW85B phototransistor (with a transient time of approximately 98μs and fall time of about 60μs). It is thus proved that the photodiode is more suitable to be used in this research compared to the phototransistors due to its fast transient and fall time.

6.1.2 Microcontroller Controlling Signals

Basically, the microcontroller remains in idle state ('0' state) until the PC sends a signal to request the DAS to start acquire data ('1' state). When the micro-controller is activated by the input signal, it will produce signals according to the programmed pulses. The PIC16F84A is programmed for two different modes for both the 2-projection mode and the 4-projection mode. In the 2-projection mode, the decoder requires 16 pulses to operate while the 4-projection mode needs only 8 pulses to function as shown in Figure 22 and Figure 23.

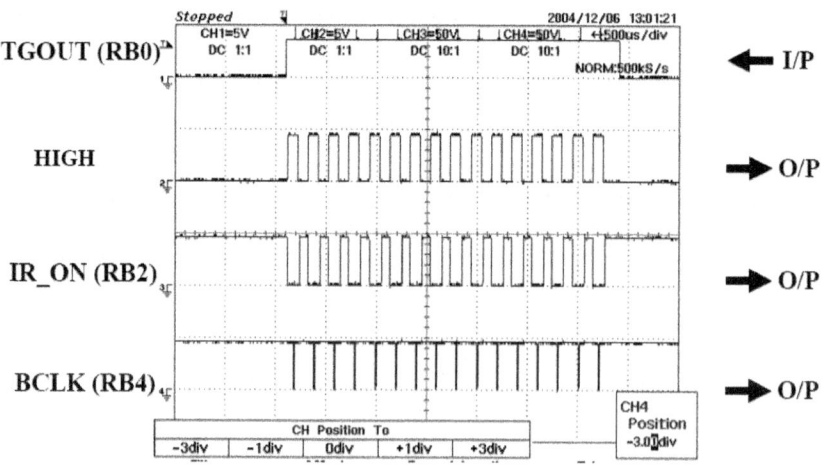

Figure 22. Microcontroller output signals in 2-projection mode (16 pulses).

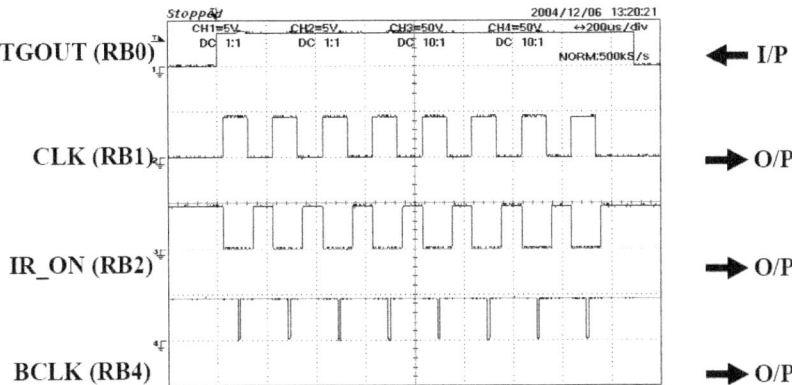

Figure 23. Microcontroller output signals in 4-projection mode (8 pulses).

The CLK signal is the 'heartbeat' to the other control signals which is set at 5kHz in this research. The first rising edge of CLK will supply an 'on' pulse for emitter to start emitting light, as shown in Figure 24. As there are 32 transmitters used in two types of projection modes, the light sequence for one frame of light emission is tabulated in Table 1.

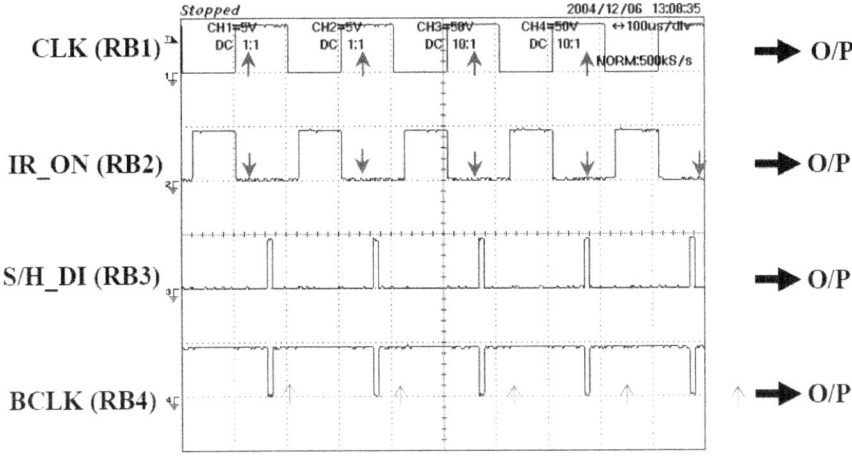

Figure 24. Timing and output control signals.

At the positive edge of the CLK signal too, the IR_ON will supply a negative edge trigger to the $\overline{G1}$ of the 74HC154 decoder (please refer to Figure 11) since $\overline{G1}$ is an active low pin. The decoder will stay activated every time the IR_ON signal is '0' and after that deactivated when signal is '1.'

Meanwhile, the rising edge of S/H_DI will set S/H signal to '1' to sample all output signals for a short period of 10μs and then continue to hold the sampled output signals until it receives the next rising edge. At the same time when the S/H signals are on hold, the BCLK signal will send a positive edge signal to the DAS to start data conversion. The total duration of the data conversion time depends on the burst mode clock frequency of 333 kHz in for this DAS. At the minimum sampling time of 3ms, the ideal conversion time would be 96ms; however due to the delays occurring while sending data to the DAS and the practical sampling time of 5ms, the conversion time is set at 390ms to ensure all data are converted properly.

Table 1. Light sequence for transmitters in one frame.

CLK Pulse Number	Tx Group (2-projection mode)	Tx Group (4-projection mode)
0	Tx0, Tx16	Tx0, Tx16, Tx8, Tx24
1	Tx1, Tx17	Tx1, Tx17, Tx9, Tx25
2	Tx2, Tx18	Tx2, Tx18, Tx10, Tx26
3	Tx3, Tx19	Tx3, Tx19, Tx11, Tx27
4	Tx4, Tx20	Tx4, Tx20, Tx12, Tx28
5	Tx5, Tx21	Tx5, Tx21, Tx13, Tx29
6	Tx6, Tx22	Tx6, Tx22, Tx14, Tx30
7	Tx7, Tx23	Tx7, Tx23, Tx15, Tx31
8	Tx8, Tx24	N/A
9	Tx9, Tx25	N/A
10	Tx10, Tx26	N/A
11	Tx11, Tx27	N/A
12	Tx12, Tx28	N/A
13	Tx13, Tx29	N/A
14	Tx14, Tx30	N/A
15	Tx15, Tx31	N/A

6.1.3 Output Voltages

There are two levels of signal conditioning circuit, which are the pre-amp stage and the amplification stage. The output of the first stage usually consists of weak signals in the range of micro volts. These low level signals are then amplified with a certain gain until they are in the suitable range required for data conversion. The amplified output voltages will then be sent to the sample and hold. The digital control input from the microcontroller S/H_DI will drive the S/H to sample the waveform and then hold the sampled signal. These signals are then sent to the DAS for conversion. As an example, Figure 25 shows the preamp voltages, amplified voltages and sampled signals of Rx23 as an object passes through the sensing beam.

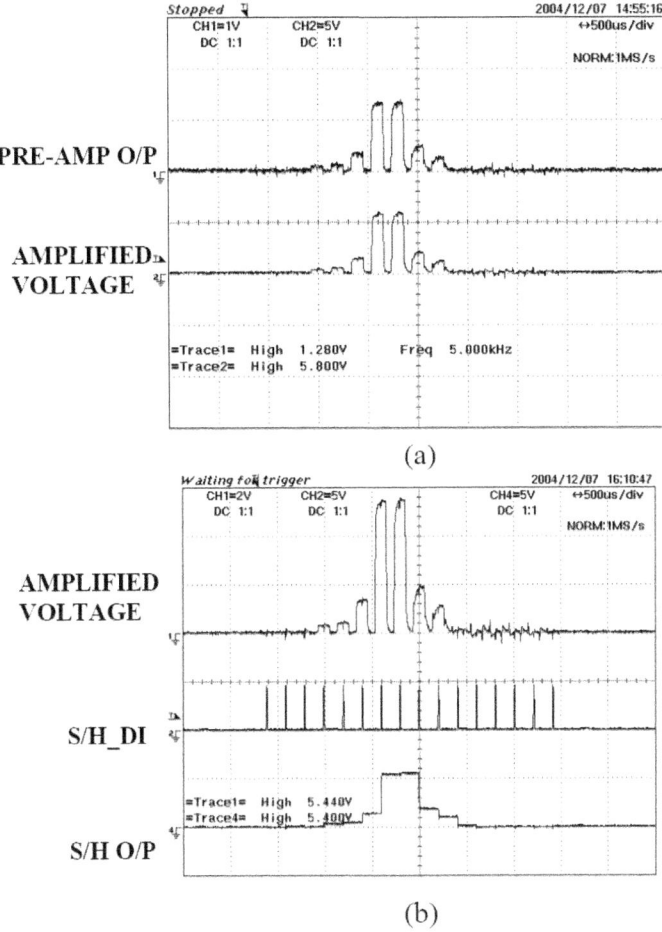

Figure 25. Output voltages for Rx23 in various stages **(a)** Pre-amp and amplification output signals **(b)** Amplified output signals, S/H_DI signal and output signals for S/H.

6.1.4 Data Acquisition Rate (DAR)

The Data Acquisition Rate or DAR can be defined as the measurement of how fast the acquired signals are transferred from the hardware to the DAS in one frame. Basically, it can be explained in a simple manner according to Equation 5.

$$DAR = \frac{1}{Total\ Conversion\ Time} \tag{5}$$

Whereby:

DAR = data acquisition rate in frames per second (unit fps).

$Total\ Conversion\ Time$ = the total time needed to convert all the 32 receivers' signals in one frame (either in 2-projection or 4-projection mode).

The rising edge of TGOUT signal is generated from the DAS when user sends a signal to the DAS to start conversion. It remains at 5 volt until one frame of conversion process finishes. Thus, if we probe the TGOUT signal, we can measure the total conversion time for one frame of data. For a system which runs at 5 kHz in this research, the TGOUT signals probed for both the 2-projection and 4-projection modes are shown in Figure 26.

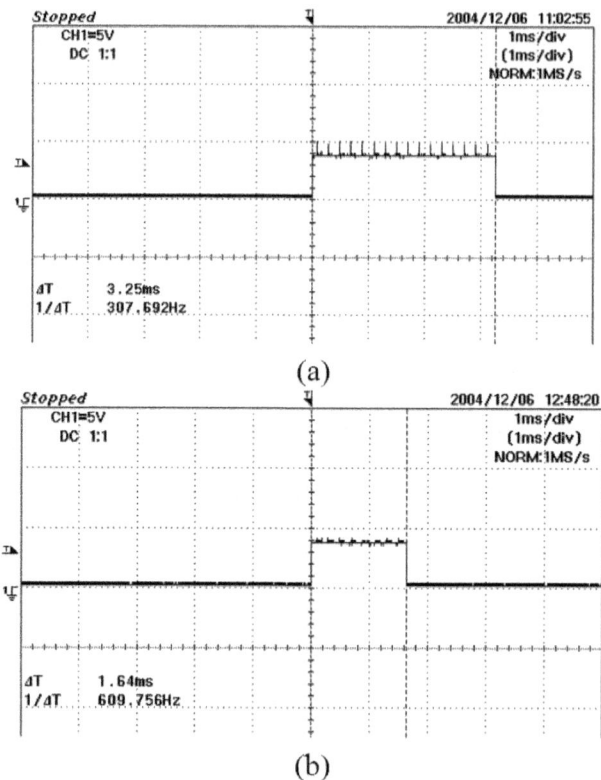

(a)

(b)

Figure 26. Total conversion time for one frame data **(a)** 2-projection **(b)** 4-projection.

Based on Equation 6.1, the DAR obtained for both the projection modes are shown in Table 2.

Table 2. DAR for different projection modes.

Projection Mode	Total Conversion time	DAR
2-projection	3.25 ms	307.69 fps
4-projection	1.64 ms	609.76 fps

It is proven here that the 4-projection mode has the ability to achieve higher DAR compared to the 2-projection mode. In the previous optical fan beam tomography research by Abdul Rahim [26], he used a total of 16 receivers with single projection each for 16 transmitters. He has managed to achieve a DAR of 300 fps. Theoretically, by using the conventional single projection technique with an increased number of sensors, the total time to convert one frame of data would be longer. It is known that a high DAR when acquiring data is essential in optical tomography system to prevent data loss.

Thus, by comparing the number of sensors and DAR obtained by Chan [7] with the results achieved in this research, it has been verified that the multiple projection technique has a capability to increase the resolution of the hardware system (a higher number of sensors installed) and at the same time increasing the DAR (shorter time needed for data conversion in one frame). The graph shown in Figure 27 represents the improvement for the DAR achieved by multiple projection technique in this research when compared to the single projection result achieved by Chan [10].

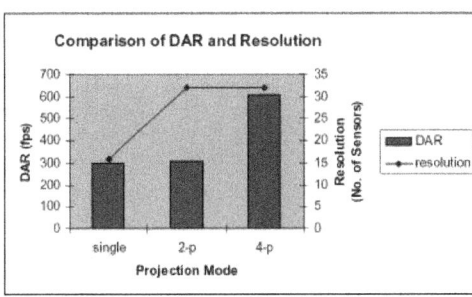

Figure 27. Comparison of DAR and resolution.

In the graph, the resolution represents the number of sensors installed in the hardware system. The 2-projection technique spots an increase of 2.56% while the 4-projection technique shows an increase of about 103.25% in DAR compared to the previous research by Abdul Rahim [13].

7. CONCLUSIONS

This paper summarizes the hardware configuration and design for this project. To design the whole hardware, it is utmost important to take note on choosing the most suitable optical sensors, preparing the fibre optics and studying on the electronics and digital systems in order to design the associated circuits. PCB drawing skill must be acquired and the PIC instruction sets must be studied to enable source code writing and programming of PIC16F84A.

REFERENCES

1. *Process Tomography: Principles, techniques and applications*; Williams, R.A.; Beck, M.S., Eds.; 1995; pp. 3-12.

2. West, R.M.; Jia, X.; Williams, R.A. Parametric Modelling in Industrial Process Tomography. In 1ˢᵗ *World Congress on Industrial Process Tomography*; 1999, pp. 444-450.

3. Ruzairi, A.R. A Tomography Imaging System for Pneumatic Conveyors Using Optical Fibres. Ph.D. Thesis, Sheffield Hallam University, 1996.

4. Sallehuddin I. Measurement of Gas Bubbles in a Vertical Water Column Using Optical Tomography. Ph.D. Thesis, Sheffield Hallam University, **2000**.

5. Khoo, B.F. Optical Fibre Sensors for Process Tomography. B.Sc. Thesis, Universiti Teknologi Malaysia, **2002**.

6. Hisyamuddin, S. Sistem Tomografi Optik Berkejituan Tinggi. B.Sc. Thesis, Universiti Teknologi Malaysia, **2001**.

INDEX